United Arab Emirates Keys to Soil Taxonomy

Shabbir A. Shahid • Mahmoud A. Abdelfattah
Michael A. Wilson • John A. Kelley
Joseph V. Chiaretti

United Arab Emirates Keys to Soil Taxonomy

Springer

Shabbir A. Shahid
Soil Correlator/Lead Soil Taxonomist
Soil Management Program
Research and Innovations Division
International Center for Biosaline Agriculture
Dubai, UAE

Michael A. Wilson
Research Soil Scientist, USDA-NRCS
National Soil Survey Center
Soil Survey Research and Laboratory
Lincoln, NE 68508-3866, USA

Joseph V. Chiaretti
Soil Scientist, USDA-NRCS
National Soil Survey Center
Soil Survey Standards Branch
Lincoln, NE 68508-3866, USA

Mahmoud A. Abdelfattah
Soil Scientist, Soil Quality Department
Environment Agency - Abu Dhabi
Abu Dhabi, UAE

Associate Professor
Soils and Water Sciences Department
Faculty of Agriculture, Fayoum University
Fayoum, Egypt

John A. Kelley
Regional Soil Scientist, retired
USDA-NRCS
Raleigh, NC 27613, USA

ISBN 978-94-007-7419-3 ISBN 978-94-007-7420-9 (eBook)
DOI 10.1007/978-94-007-7420-9
Springer Dordrecht Heidelberg New York London

Library of Congress Control Number: 2013956972

Printed on acid-free paper

Springer is part of Springer Science+Business Media (www.springer.com)

Federal laws protect wildlife (Sand gazelle-gazelle subgutturosa) in the UAE deserts

Wetlands in the desert environment are attraction to Flamingo (Phoenicopterus ruber)

From ICBA Director General

The *United Arab Emirates Keys to Soil Taxonomy* is a cooperative effort of ICBA, EAD, and USDA-NRCS staff. This cooperation has been recognized by J. Cameron Loerch (National Leader, Soil Survey Standards, USDA-NRCS) which sets a way forward of cooperation between these organizations for many years to come. The development of this valuable piece of work is unique in the Arab region and the authors deserve special congratulations. The book specifically addresses soil taxonomy in the UAE; however, it is equally good for use in the Gulf States where Aridisols and Entisols are the major soil orders. I believe the soil researchers in future will be significantly benefitted from this valuable resource in their endeavor to assess soil resources in the UAE and the Gulf Region.

Dr. Ismahane A. Elouafi
Director General
International Center for Biosaline Agriculture, Dubai UAE

Proper management of sandy soils (Entisols) is a pre-requiste for desert farming

From EAD Secretary General

The *United Arab Emirates Keys to Soil Taxonomy* is an essential tool for further soil classification studies in the United Arab Emirates. The challenge of protecting and managing the environment is significant and must be based on robust scientific and technical knowledge. Knowledge and understanding of soil is critical, particularly in an arid environment, as the resources in soil are scarce and must answer competing demands from agriculture to urban development, mineral exploration, and infrastructure development. In this regard, many approaches have been developed to partition the national landscapes into meaningful soil units from classification and land use perspectives. This book will facilitate national land use management and planning and will be equally useful for agricultural graduates studying soil science. The authors are congratulated for this unique achievement.

Razan K. Al Mubarak
Secretary General
Environment Agency—Abu Dhabi

Flooded interdunal sabkha is common in the south of UAE, high dunes in the background is dominant landscape

From USDA-NRCS National Leader

The *United Arab Emirates Keys to Soil Taxonomy* provides a basic soil classification system that will prove to be very beneficial to anyone wishing to classify and correlate soils within the region. It not only provides the taxonomic key necessary for the classification of soils that can be easily used in the field, but establishes a scientific standard to be used (and updated as needed) for future generations. The authors are congratulated for this outstanding collaborative effort in developing a dynamic system that will serve current and future generations of soil scientists as they strive to better understand and educate the public about this vital natural resource *the soil.*

J. Cameron Loerch
National Leader, Soil Survey Standards
USDA-Natural Resources Conservation Service

Diversified landscape showing sandy soils and wetland in mountainous area of UAE

About the Associate Authors

Collectively from ICBA, EAD, and USDA-NRCS, the authors have brought a vast range of knowledge and experience together in the development of the *United Arab Emirates Keys to Soil Taxonomy*.

Dr. Shabbir A. Shahid has more than 32 years of experience as a Soil Scientist in Pakistan, UK, Australia, Kuwait, and the United Arab Emirates. He served as Lead Soil Taxonomist, Technical Coordinator, and QA Expert for the Soil Surveys of Kuwait (1996–1999) and Abu Dhabi Emirate (2004–2009). He with his co-associates discovered anhydrite soil in the coastal land of Abu Dhabi Emirate. He is a prolific author and coauthor of over 150 scientific papers published in peer-reviewed scientific journals, books, and manuals. He pioneered the concept of soil survey in the UAE.

E-mail: s.shahid@biosaline.org.ae

Dr. Mahmoud A. Abdelfattah served as a Mapping Crew Leader and Deputy Technical Coordinator for the Soil Survey of the Abu Dhabi Project (2004–2009) and Technical Manager for the Soil Survey of the Northern Emirates (2010–2012). He has over 20 years of experience in teaching and research. He has participated in a number of international conferences, managed a number of technical projects, and has authored and coauthored over 50 published scientific papers and book chapters.

E-mail: mabdelfattah@ead.ae

Dr. Michael A. Wilson is a Research Soil Scientist at the USDA-NRCS National Soil Survey Center in Lincoln, Nebraska. He has served in that position for over 20 years, conducting soil genesis research, specifically in the areas of soil geochemistry and mineralogy. He was part of a review team for the Abu Dhabi Soil Survey and has conducted research on the quantification and distribution of anhydrite in the coastal sabkha soils of Abu Dhabi.

E-mail: mike.wilson@lin.usda.gov

Mr. John A. Kelley has more than 33 years of experience as a USDA Soil Scientist and Regional Correlator in the eastern USA and served as Technical Coordinator and Quality Assurance Specialist for the Soil Survey of the Northern Emirates. He has contributed in many ways toward the completion of this key, serving as Technical Editor and helping with photography including photo editing and image processing.

E-mail: bettmark.john@gmail.com

Mr. Joseph V. Chiaretti is a Soil Scientist with the USDA and serves on the soil survey standards staff at the NRCS National Soil Survey Center in Lincoln, Nebraska. He is responsible for developing and maintaining soil survey division handbooks and other technical documents. Prior to his current position, he has had 11 years of experience in conducting quality assurance on soil survey data and 19 years of experience in leading and field mapping on soil survey projects in the southwestern USA.

E-mail: joe.chiaretti@lin.usda.gov

Indepth exploration of a soil taxa in the coastal land of UAE

Preface

All terrestrial life ultimately depends on soil, energy, and water. Soils have always been central to human civilization and life. They are an integral part of the physical and cultural environment, and we may take them for granted and even tend to treat them contemptuously in the United Arab Emirates (UAE). The rise and fall of civilizations have been closely linked with the use and abuse of soil and water resources. There is little reason to believe that these linkages will disappear in the future. It is therefore important to evaluate soils for their quality and link them to appropriate uses and services. In this publication, information is provided on soil classification and how to key out taxa relevant to UAE soils.

The recent soil inventory of the United Arab Emirates revealed that the UAE landscape is covered mainly by low-lying sandy deserts, mega-barchan dunes, extensive coastal salt flats, and alluvial and gravelly plains in both the far west and the east. The recent soil surveys revealed that a rather uniform-looking desert landscape has, in fact, a diversity of subsurface features that help to categorize the soils into 74 soil series and their phases. These features confirm the soil diversity in terms of classification, chemistry, physics, mineralogy, fertility, suitability for different uses, and vulnerability to land degradation.

The objectives of this book are to provide information for keying out the soils of the United Arab Emirates into separate classes and to provide a guide to associated laboratory methods. The classification used predominantly is extracted from the 11th edition of the USDA-NRCS *Keys to Soil Taxonomy*, and sections relevant to the soils found in the UAE are included here.

Primarily, this key is designed to fit the soil system of the United Arab Emirates. Information not found in the USDA key has been added including criteria and classes for: (1) differentiating anhydritic soils from gypsic soils; (2) identifying "lithic" subgroups for Aquisalids and Haplosalids; (3) identifying "salidic" subgroups within the great groups of Gypsids, Calcids, Psamments, and Orthents; and (4) incorporation of phases for soil taxa. The classes for the newly identified anhydrite soils in the UAE have been added at four different levels: the *anhydritic* subsurface diagnostic horizon and mineralogy class and the *Anhydritic Haplosalids* and *Anhydritic Aquisalids* subgroups. In addition, a horizon suffix of "aa" for layers with an accumulation of anhydrite has been incorporated. The concept of horizon suffix "k" also has an ad hoc expansion, beyond the official definition of pedogenic accumulations, to connote the simple presence of calcium carbonate as determined by effervescence in dilute hydrochloric acid. This usage is synonymous with the recently defined soil characteristic named "free carbonates" in the *Keys to Soil Taxonomy*. The added classes or features that are proposed for USDA *Soil Taxonomy* are designated with a "†" in the Table of Contents and are footnoted in the text. The classes are in different stages of review and approval for use in the USDA soil taxonomy system; however, discussions regarding final approval and incorporation of the additions are ongoing. Other additions such as "Phases of soil taxa" are unique to the *United Arab Emirates Keys to Soil Taxonomy* and are not proposed for addition to the USDA system.

This book will provide a mechanism for updating the current soil surveys and will facilitate the correlation of soils from new surveys within the UAE. Additionally, this book provides a source of information to help the international soil science community converse about UAE soils and their comparison to other soils. Commonality between classification systems used in different countries enhances linkages. These linkages allow countries with similar mapping and classification procedures and similar soils to transfer agriculture technology without conducting long-term experiments under similar environmental conditions.

We hope the countless number of users of soil surveys in the UAE and abroad will use this publication to learn about classification of the soils of the United Arab Emirates.

Shabbir A. Shahid
Team Leader, International Center for Biosaline Agriculture
Dubai, UAE

Surface gravel layer (armor layer) is an indicator of land degradation

Camel grazing is common in high sand dune area

Acknowledgements

The authors would like to thank both the management of the Dubai-based International Center for Biosaline Agriculture and the Environment Agency—Abu Dhabi for their partnership in the joint completion of the state-of-the-art Soil Survey of Abu Dhabi Emirate and preparation of the action plan for the Soil Survey of the Northern Emirates. The management of the UAE Ministry of Environment and Water and EAD deserve special thanks for the joint implementation of the Northern Emirates Soil Survey project. Special thanks to H.E. Mr. Majid Al Mansouri, former Secretary General of the Environment Agency—Abu Dhabi, H.E. Ms. Razan Khalifa Al Mubarak, EAD Secretary General, and H.E. Dr. Jaber Al Jaberi, EAD Deputy Secretary General, for their full support during the implementation of the Soil Survey of Abu Dhabi and the Soil Survey of the Northern Emirates, which this key is based upon.

The Australian contractor GRM International deserves special thanks for the completion of both surveys. The soil survey reports were used as the basis for the compilation of the *United Arab Emirates Keys to Soil Taxonomy.*

Special thanks go to Dr. Ismahane Elouafi (Director General ICBA) and Dr. Ahmed A. Al Sharif (Deputy Director General ICBA) for their encouragement and interest in completing this important publication that will be used by UAE nationals for many years ahead.

Associate Authors

Soil Scientists from ICBA, EAD, USDA-NRCS and CSIRO jointly visited newly discovered Anhydrite soil in Abu Dhabi Coastal land

Contents

Small micronebkhas (wind-blown sand mounds with Zygophyllum qatarense) protects wind erosion of psamments

Interdunal sabkha (Salids) are common in the south of UAE

List of Figures

List of Tables

Investigation of the redoximorphic features in a soil from Abu Dhabi coastline

Soil Classification

1

Abstract

The global knowledge of soil classification is continually advancing; therefore, soil taxonomies are regularly upgraded as the new soils are identified. Presently, several soil classification systems are used worldwide. Most systems in use generally provide guidelines to key out soils within their national boundaries. Some examples are Soil Classification for England and Wales, the Soil Map Legend of the World, World Reference Base for Soil Resources, the Canadian System of Soil Classification, the Australian Soil Classification, the New Zealand Soil Classification and the Russian Soil Classification System. The US Soil Taxonomy is a comprehensive system to be used globally, and has been used extensively in over 75 countries. No one country has within its boundaries all the soils identified in any soils key: therefore the user in the UAE has to go through all the list of potential soils to key out a specific soil in the study area. This issue suggests a need to develop Keys to Soil Taxonomy addressing the soils within the national boundaries of any country; for example the UAE. The United Arab Emirates Keys to Soil Taxonomy is a collection of information from UAE soil surveys and is specifically useful at the national level, or for other Gulf Cooperation Council countries (Bahrain, Kuwait, Oman, Qatar, Saudi Arabia) where these soils have been mapped in national soil inventories.

Keywords

Soil classification systems • Soil map • Soil taxonomy • UAE

S.A. Shahid et al., *United Arab Emirates Keys to Soil Taxonomy*,
DOI 10.1007/978-94-007-7420-9_1,

1.1 Introduction

Soil classification systematically arranges soils into groups or categories on the basis of their characteristics, whereas the soil survey is a systematic examination, description, classification, and mapping of soils in an area. The global knowledge of soil classification is continually advancing; therefore, soil taxonomies are continuously upgraded. Presently, several soil classification systems are used worldwide. Among these, *Soil Taxonomy* (Soil Survey Staff 1999) is very comprehensive and has been used extensively, including GCC countries (Saudi Arabia, Oman, Kuwait, Qatar, Abu Dhabi Emirate, Dubai Emirate, and the Northern Emirates of UAE). Other systems in use are Soil Classification for England and Wales (Avery 1980), the Soil Map Legend of the World (FAO-UNESCO 1990), World Reference Base for Soil Resources (WRB) (IUSS Working Group WRB 2006), the Canadian System of Soil Classification (Canadian Soil Survey Committee 1978), the Australian Soil Classification (Isbell 1998), the New Zealand Soil Classification (Hewitt 1992), and the Russian Soil Classification System (Shishov et al. 2004).

In the Dubai Emirate, soil taxonomy was used to classify soils at 1:25,000 scale at soil series level (DM 2005). In the Abu Dhabi Emirate, soil taxonomy was used to classify soils at two survey levels: (1) Extensive survey—a fourth order level modified to fit Abu Dhabi soil conditions at a scale of 1:100,000; and (2) Intensive survey—a second order level modified to fit Abu Dhabi soil conditions at a scale of 1:25,000 (EAD 2009a, b). The complete report consists of five volumes, including an Executive Summary. The soil survey of the Northern Emirates (EAD 2012) is a third order level soil survey conducted at a scale of 1:50,000. The complete report consists of three volumes including an Executive Summary.

The focus of this publication is to provide useful information to soil surveyors in the UAE, as well as other soil scientists around the world, to key out soils of the Emirates. The information provided in this book is principally based on the Dubai Emirate, Abu Dhabi and Northern Emirates soil surveys. However, in the future, more taxa will likely be added to this book as mapping and research recognize additional soils in the UAE. Thus, soil scientists and other users should classify the soils using the most recent version of *The United Arab Emirates Keys to Soil Taxonomy*.

References

Avery, B. W. (1980). *Soil survey – soil classification for England and Wales (Higher categories)*. Technical Monograph No. 14. Soil Survey, Rothamsted Experimental Station (p. 67). Lawes Agricultural Trust, Harpenden Herts AL5 2JQ UK.

Canadian Soil Survey Committee. (1978). *The Canadian system of soil classification* (Research Branch Publication, No. 1646). Ottawa: Canada Department of Agriculture.

Dubai Municipality (DM). (2005). *Satellite imagery and thematic mapping project*, (p. 221). Executed jointly by Global Scan Technologies (Dubai) and National Remote Sensing Agency, India.

EAD. (2009a). *Soil survey of Abu Dhabi Emirate-extensive survey* (Vol. 1, pp. xx+506). Environment Agency – Abu Dhabi, UAE.

EAD. (2009b). *Soil survey of Abu Dhabi Emirate-intensive survey* (Vol. 2, pp. xvii+435). Environment Agency – Abu Dhabi, UAE.

EAD. (2012). *Soil survey of the Northern Emirates.* (Soil Report, Vol. 1, pp. xxvi+410). Environment Agency – Abu Dhabi, UAE.

FAO-UNESCO. (1990). *Soil map of the world: Revised legend* (World Soil Report, No. 60). Rome: FAO.

Hewitt, A.A. (1992). *New Zealand soil classification.* DSRI Land Resources Scientific Report No. 19.

Isbell, R. F. (1998). *The Australian soil classification.* Collingwood: CSIRO.

IUSS Working Group, WRB. (2006). *World reference base for soil resources* (World Soil Report, No. 103). Rome: FAO.

Shishov, L. L., Tonkonogov, V. D., Lebedeva, I. I., & Gerasimova, M. I. (2004). *Classification and diagnostics of Russian soils.* Smolensk: Oikumena [in Russian].

Soil Survey Staff. (1999). *Soil taxonomy: A basic system of soil classification for making and interpreting soil surveys* (U.S. Department of Agriculture Handbook 2nd ed., No. 436). Washington, DC: U.S. Government Printing Office.

Agriculture farming on an interdunal plain at the base of megadunes

The Soils That We Classify

2

Abstract

Soil is the natural medium for the growth of land plants, whether or not it has discernible soil horizons. People consider soil important because it supports plants that supply food, fibers, drugs, and other human needs and because it filters water and recycles wastes. Soil is a natural body comprised of solids (minerals and organic matter), liquid, and gases that occur on the land surface, occupies space, and is characterized by horizons, or layers, that are distinguishable from initial geologic material as a result of additions, losses, transfers, and transformations of energy and matter or the ability to support rooted plants in a natural environment. The upper limit of soil is the boundary between soil and air, shallow water, live plants, or plant materials that have not begun to decompose. For purposes of classification, the lower boundary of soil is arbitrarily set at 200 cm. In areas where soil has thin cemented horizons that are impermeable to roots, the soil extends as deep as the deepest cemented horizon, but not below 200 cm.

The United Arab Emirates Keys to Soil Taxonomy is designed in such a way that the reader makes the correct classification by going through the key systematically, starting at the beginning and eliminating one by one all classes that include criteria that do not fit the soil or layer in question.

Keywords

Cemented horizons • Geological material • Land plants • Natural body • Lower boundary

S.A. Shahid et al., *United Arab Emirates Keys to Soil Taxonomy*,
DOI 10.1007/978-94-007-7420-9_2,

2.1 Introduction

This section in general is taken from *Keys to Soil Taxonomy* (Soil Survey Staff 2010) with slight modifications that reflect the soil conditions and the soil system in the United Arab Emirates.

Soil is the natural medium for the growth of land plants, whether or not it has discernible soil horizons. This meaning is still the common understanding of the term, and the greatest interest in soil is centered on this meaning. People consider soil important because it supports plants that supply food, fibers, drugs, and other wants of humans and because it filters water and recycles wastes. Soil is a natural body comprised of solids (minerals and organic matter), liquid, and gases that occur on the land surface, occupies space, and is characterized by one or both of the following: horizons, or layers, that are distinguishable from initial geologic material as a result of additions, losses, transfers, and transformations of energy and matter *or* the ability to support rooted plants in a natural environment. This definition is expanded from the 1975 version of *Soil Taxonomy* (USDA, SCS 1975) to include soils in areas of Antarctica where pedogenesis occurs but where the climate is too harsh to support the higher plant forms.

The upper limit of soil is the boundary between soil and air, shallow water, live plants, or plant materials that have not begun to decompose. Areas are not considered to have soil if the surface is permanently covered by water too deep (typically more than about 2.5 m) for the growth of rooted plants. The lower horizontal boundaries of soil are areas where the soil grades to deep water, barren areas, rock, or ice. In some places the separation between soil and nonsoil is so gradual that clear distinctions cannot be made.

The lower boundary that separates soil from the nonsoil underneath is most difficult to define. Soil consists of the horizons near the earth's surface that, in contrast to the underlying parent material, have been altered by the interactions of climate, relief, and living organisms over time. Commonly, soil grades at its lower boundary to hard rock or to earthy materials virtually devoid of animals, roots, or other marks of biological activity. The lowest depth of biological activity, however, is difficult to discern and is often gradual. For purposes of classification, the lower boundary of soil is arbitrarily set at 200 cm. In soils where either biological activity or current pedogenic processes extend to depths greater than 200 cm, the lower limit of the soil for classification purposes is still 200 cm (Fig. 2.1). In some instances the more weakly cemented bedrocks (paralithic materials) have been described and used to differentiate soil series (series control section), even though the paralithic materials below a paralithic contact are not considered soil in the true sense. In areas where soil has thin cemented horizons that are impermeable to roots, the soil extends as

Fig. 2.1 Soil scientists examining and describing a typical soil profile. The scientists document several soil properties and features, commonly to a depth of 2 m. Note the distinct change in horizons (or layers) with depth

deep as the deepest cemented horizon, but not below 200 cm. For certain management goals, layers deeper than the lower boundary of the soil that is classified (200 cm) must also be described if they affect the content and movement of water and air or other interpretative concerns.

2.2 Buried Soils

A buried soil is covered with a surface mantle of new soil material that either is 50 cm or more thick or is 30–50 cm thick and has a thickness that equals at least half the total thickness of the named diagnostic horizons that are preserved in the buried soil. A surface mantle of new material that does not have the required thickness for buried soils can be used to establish a map unit phase of the mantled soil or even another soil series if the mantle affects the use of the soil.

A surface mantle of new material, as defined here, is largely unaltered, at least in the lower part. It may have a diagnostic surface horizon (epipedon) and/or a cambic horizon, but it has no other diagnostic subsurface horizons, all defined later. However, there remains a layer 7.5 cm or more thick that fails the requirements for all diagnostic horizons, as defined later, overlying a horizon sequence that can be clearly identified as the solum of a buried soil in at least half of each pedon. The recognition of a surface mantle should not be based only on studies of associated soils. Soil, as defined in this text, does not need to have discernible horizons, although the presence or absence of horizons and their nature are of extreme importance in soil classification. Plants can be grown under glass in pots filled with earthy materials, such as peat or sand, or even in water. Under proper conditions all these media are productive for plants, but they are nonsoil here in the sense that they cannot be classified in the same system that is used for the soils of a survey area, county, or even a nation. Plants even grow on trees, but trees are regarded as nonsoil.

Soils are three dimensional (3-D) bodies, and many scientists have different opinions about the classification. From a soil classification point of view, however, the entity that is classified is the *soil profile*, which is a vertical section of soil through different horizon to the parent material. Soil has many properties that fluctuate with the seasons. It may be alternately cold and warm or dry and moist. Biological activity is slowed or stopped if the soil becomes too cold or too dry. The soil receives flushes of organic matter when leaves fall or grasses die. Soil is not static. The pH, soluble salts, amount of organic matter and carbon-nitrogen ratio, numbers of micro-organisms, soil fauna, temperature, and moisture all change with the seasons as well as with more extended periods of time. Soil must be viewed from both the short-term and long-term perspective.

Soils are (Shahid 2007):

- The essence of life
- A product of the environment
- Developed, and not merely an accumulation of debris from rocks and organic matter
- Different from the material from which they are derive
- Continuously changing due to natural and human influences
- Different in inherent capability
- The sites for chemical reactions and organisms
- A medium for the support of plants
- A medium for filtering water and recycling wastes
- Very fragile
- Very slowly renewable

The hierarchy of Dubai (DM 2005), Abu Dhabi (EAD 2009a, b; Shahid et al. 2004) and Northern Emirates soil classification (EAD 2012) has been used in the preparation of this book to help the new user assign soil taxonomic classifications in areas of his/her interest for the maintenance of soil surveys in the UAE. It is most likely that in the process of future soil surveys, new classifications will be identified. In the light of new findings, this edition will be updated periodically as needed.

2.3 Approach to Soil Classification

When classifying soils the user must first consider: (1) the depth in which the soil feature (property or characteristic) occurs; (2) the thickness of the layer where it occurs; and (3) the extent or degree of expression that the feature exhibits. Those soil properties that have a major influence for use and management of the soil are defined as diagnostic horizons. For the diagnostic subsurface horizons identified in the UAE to be recognized at the highest levels of classification (order, suborder, great group, or subgroup) they must occur within a depth of 100 cm from the soil surface. This is the predominant rooting depth of most plants commonly grown. If these features are present but occur at a lower depth (but within 200 cm) they still play an important role in land use planning and have been identified as phases.

Often the soil feature critical to classification is confirmed by simple observation (Fig. 2.2) and volume estimation (qualitative analysis). However, once the feature has been identified, laboratory support is commonly needed to quantify the amount or degree of expression to confirm classification. For example, the presence of calcium carbonate may be observed in the soil profile in the form of coatings, masses, or nodules and a subjective estimate (by volume) is made. However, to confirm the presence of a calcic horizon samples must be collected and analyzed and the presence of calcium carbonate determined quantitatively (by weight). The amount of the material, depth in which the material was observed, and the thickness of the horizon in which it occurs all go together to document the presence of the diagnostic horizon and subsequent taxonomic placement.

The *United Arab Emirates Keys to Soil Taxonomy* is designed in such a way that the reader makes the correct classification by going through the key systematically, starting at the beginning and eliminating one by one all classes that include criteria that do not fit the soil or layer in question. The key has been designed to provide separation of major soil features. In many sections the sequence of the key is based on the severity of limitation the particular property may pose.

Fig. 2.2 Any information collected at a soil description site is carefully documented and geo-referenced

2.4 Naming Subgroups

Subgroups are keyed in a specific and consistent order based on their limitation to use and management of the soils. Subgroups are arrayed in order of most limiting to least limiting. For example, within any great group the Lithic subgroup always keys first. The significant limitation posed by shallow depth to hard bedrock is considered the most limiting of all subgroups to land use and management.

Subgroup names are applied in the following order:

Lithic
Petrogypsic
Petrocalcic
Anhydritic
Gypsic
Leptic

Calcic
Salidic
Sodic

If none of the listed subgroups apply, the subgroup is considered *typic* or typical for the central concept of the great group.

Many soil taxa have properties that encompass more than one subgroup. If more than one subgroup is needed, additional names are added. To date, only two subgroups modifiers have been recognized for individual great groups identified in the United Arab Emirates. The most limiting subgroup is used to initially modify the great group and additional names are added in order of significance. For example, a pedon classified in a Torripsamment great group has properties of a lithic and salidic subgroup. The pedon would be classified at the subgroup level as a "Salidic Lithic Torripsamment". Currently, it is recommended to identify more than two subgroups as differentiae at the series level and the classification remain unchanged.

References

Dubai Municipality (DM). (2005). *Satellite imagery and thematic mapping project* (p. 221). Executed jointly by Global Scan Technologies (Dubai) and National Remote Sensing Agency, India.

EAD. (2009a). *Soil survey of Abu Dhabi Emirate—extensive survey* (Vol. 1, pp. xx+506). Environment Agency - Abu Dhabi, UAE.

EAD. (2009b). *Soil survey of Abu Dhabi Emirate-intensive survey* (Vol. 2, pp. xvii+435). Environment Agency - Abu Dhabi, UAE.

EAD. (2012). *Soil survey of the Northern Emirates* (Soil Report, Vol. 1, pp. xxvi+410). Environment Agency - Abu Dhabi, UAE.

Shahid, S.A. (2007). Soils. In *Physical geography*, Abu Dhabi Global Environmental Data Initiatives (pp. 6–41). Environment Agency - Abu Dhabi, UAE.

Shahid, S. A., Abdelfattah, M. A., Arshad, K. R., Muhairbi, M., Al Othman, Y. R., Al Haji, A., Mansouri, M., Al Ahmad, S., Siddiqui, A., Azimullah, K. M., Sekaly, A. L., & Al Qubaisi, B. S. (2004). *Soil survey for the coastline of Abu Dhabi Emirate. Two volumes (Vol. 1: Reconnaissance Survey; Vol. 2: Soil Maps).* Abu Dhabi: Unpublished report of the Environmental Research and Wildlife Development Agency.

Soil Survey Staff. (2010). *Keys to soil taxonomy* (11th ed.). Washington, DC: U.S. Department of Agriculture, Natural Resources Conservation Service, U.S. Government Printing Office.

USDA, SCS. (1975). *Soil taxonomy: A basic system of soil classification for making and interpreting soil surveys* (U.S. Department of Agriculture Handbook No. 436). Washington, DC: U.S. Government Printing Office.

Multicolored soil profile indicates phases of soil development

Uniform looking desert landscapes present indepth picturesque features

Horizons, Layers, and Characteristics Diagnostic for the Higher Categories

3

Abstract

This chapter defines the horizons, layers, and characteristics of mineral soils found in the UAE. Soil profiles are described according to the different soil layers that may be identified within each profile. In this chapter "required characteristics" for horizons or features, are arranged as a key. Each major layer, or master horizon, is allocated one or more letter codes, each of which has a specific meaning. The uppercase letters O, L, A, E, B, C, R, M, and W are used within the USDA-NRCS system to represent the master horizons and layers of soils. These letters are the base symbols to which other characteristics are added to complete the designations. Lowercase letters are used as suffixes to designate specific kinds of master horizons and layers. The letters aa, b, c, d, g, k, m, n, p, q, r, w, y, yy, and z are the suffix symbols used for the UAE soils. The use of prime symbol, where appropriate, is placed after the capital-letter horizon B′k and the "caret" symbol (^), for example ^A_^C-Ak, are described. Diagnostic surface (Ochric) and subsurface horizons (Anhydritic, Calcic, Cambic, Gypsic, Petrocalcic, Petrogypsic, Salic), as well as moisture (Aquic, Aridic or Torric) and a temperature regime (Hyperthermic) identified in the UAE are described. The Anhydritic horizon is a new subsurface diagnostic horizon first recognized and defined in the UAE.

Keywords

Aridic • Diagnostic horizons • Suffix letters • Moisture regime • Temperature regime

S.A. Shahid et al., *United Arab Emirates Keys to Soil Taxonomy*,
DOI 10.1007/978-94-007-7420-9_3,

3.1 Introduction

This chapter defines the horizons, layers, and characteristics of mineral soils found in the United Arab Emirates. The horizons and characteristics defined below are not in a key format. The "required characteristics" for horizons or features, however, are arranged as a key. Some diagnostic horizons are mutually exclusive, and some are not.

3.2 Master Horizons and Layers

Soil profiles are described according to the different soil layers that may be identified within each profile. Each major layer, or master horizon, is allocated one or more letter codes, each of which has a specific meaning. The capital letters O, L, A, E, B, C, R, M, and W are used within the USDA-NRCS system (Soil Survey Staff 2010) to represent the master horizons and layers of soils. These letters are the base symbols to which other characteristics are added to complete the designations.

Common master horizons used for soils in arid regions are:

A horizons—mineral horizons that show an accumulation of humified organic matter or show signs of cultivation;

B horizons—mineral horizons that show evidence of illuvial concentration of silicate clay, gypsum, carbonates, iron, aluminum, humus, silica, or sesquioxides alone or in combination or have structure other than rock structure;

C horizons or layers—lack the properties of O, A, E, or B horizons and may be little affected by pedogenic processes;

R layers—strongly cemented to indurated bedrock. Granite, basalt, limestone, and sandstone are examples of bedrock designated by the letter R.

Within the UAE, the conditions are generally unsuitable for the occurrence of many pedogenic processes, and thus recognition of A, B, and C horizons can be difficult. B horizons are typically identified in the UAE through the accumulation of gypsum and/or carbonates, while C horizons typically lack this development. Due to the lack of significant amounts of organic matter, the surface layers are generally defined as A horizons. However, within the dune areas, the surface layer has been blown in from elsewhere and is typically the most recent material. In general, these deep sands lack any significant profile development, and thus most of the horizons (including the surface) have been described as C horizons. Lithified layers and bedrock are encountered and are designated as R layers.

3.3 Suffix Symbols

Lowercase letters are used as suffixes to designate specific kinds of master horizons and layers. The term “accumulation” is used in many of the definitions of such horizons to indicate that these horizons must contain more of the material in question than is presumed to have been present in the parent material.

The letters aa, b, c, d, g, k, m, n, p, q, r, w, y, yy, and z are the suffix symbols used for the UAE soils. These suffixes and their meanings are described below:

aa Accumulation of anhydrite†

This symbol is used in mineral soils to indicate an accumulation of anhydrite ($CaSO_4$). Colors associated with horizons that have suffix aa typically have hue of 5Y with chroma (dry and moist) of 1 or 2, and value of 7 or 8. Soil materials containing anhydrite have a fluid manner of failure. The symbol is commonly used in combination with suffix z as “Bzaa”.

b Buried genetic horizon

This symbol indicates identifiable buried horizons with major genetic features (diagonostic horizons) that were developed before burial. Genetic horizons may or may not have formed in the overlying material, which may be either like or unlike the assumed parent material of the buried soil.

c Concretions or nodules

This symbol indicates a significant accumulation of concretions or nodules. Cementation is required. The cementing agent commonly is iron, aluminum, or manganese. It cannot be silica, dolomite, calcite, or more soluble salts.

d Physical root restriction

This symbol indicates noncemented, root-restricting layers in naturally occurring or human-made sediments or materials. Examples are dense basal till, plowpans, and other mechanically compacted zones.

g Strong gleying

This symbol indicates either that iron has been reduced and removed during soil formation or that saturation with stagnant water has preserved it in a reduced state. Most of the affected layers have chroma of 2 or less, and many have redox concentrations. The low chroma can represent either the color of reduced iron or the color of uncoated sand and silt particles from which iron has been removed.

k Presence of calcium carbonates

This symbol indicates the presence of calcium carbonate as exhibited by effervescence or as visible accumulations that occur as carbonate filaments, coatings, masses, nodules, disseminated carbonate, or other forms. It was used with horizons that were very slightly to strongly effervescent (formation of bubbles) with application of dilute hydrochloric acid.

m ***Cementation or induration***

This symbol indicates continuous or nearly continuous cementation. It is used only for horizons that are more than 90 % cemented, although they may be fractured. The cemented layer is physically root restrictive. The predominant cementing agent (or the two dominant ones) may be indicated by adding defined letter suffixes, singly or in pairs. The horizon suffix km indicates cementation by carbonates; qm, cementation by silica; yym, cementation by gypsum; kqm, cementation by carbonates and silica; and zm, cementation by salts more soluble than gypsum.

n ***Accumulation of sodium***

This symbol indicates an accumulation of exchangeable sodium. It is used with horizons that have an exchangeable sodium percentage (ESP) of 15 or more or sodium adsorption ratio (SAR) of 13 $(\text{mmoles L}^{-1})^{0.5}$ or more.

p ***Tillage or other disturbance***

This symbol indicates a disturbance of the surface layer by mechanical means, pasturing, or similar uses.

q ***Accumulation of silica***

This symbol indicates an accumulation of secondary silica.

r ***Weathered or soft bedrock***

This symbol is used with C to indicate layers of bedrock that are moderately cemented or less cemented. Examples are weathered igneous rock and partly consolidated sandstone, siltstone, or shale. The excavation difficulty is low to high.

w ***Development of color or structure***

This symbol is used only with B horizons to indicate the development of color or structure, or both, with little or no apparent illuvial accumulation of material. It was not used to indicate a transitional horizon.

y ***Accumulation of gypsum***

This symbol indicates an accumulation of gypsum. It was used when the horizon fabric was dominated by soil particles or minerals other than gypsum. Gypsum is present in amounts that do not significantly obscure or disrupt other features of the horizon (generally, 1–50 % or more by volume; or 5–40 % percent or more by weight).

yy ***Dominance of horizon by gypsum***

This symbol indicates a horizon that is dominated by the presence of gypsum. The gypsum content may be due to an accumulation of secondary gypsum, the transformation of primary gypsum inherited from parent material, or other processes. Suffix yy is used when the horizon fabric has an abundance of gypsum (generally, 50 % or more, by volume; or 40 % or more, by weight) and pedogenic or lithologic features are obscured or disrupted by growth of gypsum crystals. Colors associated with horizons that have suffix yy typically are highly whitened, with value of 7 through 9.5 and chroma of 2 or less.

z ***Accumulation of salts more soluble than gypsum***
This symbol indicates an accumulation of salts that are more soluble than gypsum. The suffix z is commonly used when the ECe is more than 4 dS/m (slightly or more saline).

3.4 Conventions for Using Letter Suffixes

Many master horizons and layers that are symbolized by a single capital letter have one or more lowercase letter suffixes. The following rules apply:

1. Letter suffixes should directly follow the capital letter.
2. More than three suffixes are rarely used.
3. If more than one suffix is needed, the following letters, if used, are written first: d, k, r, t, and w. None of these letters are used in combination in a single horizon.
4. If more than one suffix is needed and the horizon is not buried, the following symbols, if used, are written last: aa, c, g, and m (e.g., Bkm or Bzaa).
5. If a horizon is buried, the suffix b is written last (e.g., Bkb).
6. If the above rules do not apply to certain suffixes, such as k, q, or y, the suffixes may be listed together in order of assumed dominance or they are listed alphabetically if dominance is not a concern.

A B horizon that is gleyed or has accumulations of carbonates, sodium, silica, gypsum, or salts more soluble than gypsum or residual accumulations of sesquioxides carries the appropriate symbol: g, k, n, q, y, or z.

3.5 Vertical Subdivision

Commonly, a horizon or layer identified by a single letter or a combination of letters has to be subdivided. For this purpose, Arabic numerals are added to the letters of the horizon designation. These numerals follow all the letters. Within a C horizon, for example, successive layers may be designated C1, C2, C3, etc. If the lower part is gleyed and the upper part is not gleyed, the layers may be designated C1-C2-Cg1-Cg2.

These conventions apply whatever the purpose of the subdivision. In many soils a horizon that could be identified by a single set of letters is subdivided because of the need to recognize differences in morphological features, such as structure, color, or texture. These divisions are numbered consecutively with numerals, but the numbering starts again with 1 wherever in the profile any letter of the horizon symbol changes, e.g., Bk1-Bk2-Bky1-Bky2 (not Bk1-Bk2-Bky3-Bky4). The numbering of vertical subdivisions within consecutive horizons is not interrupted at a discontinuity (indicated by a numerical prefix) if the same letter combination is used in both materials, e.g., Bk1-Bk2-2Bk3-2Bk4 (not Bk1-Bk2-2Bk1-2Bk2).

During sampling for laboratory analyses, thick soil horizons are sometimes subdivided even though differences in morphology are not evident in the field. These subdivisions are identified by numerals that follow the respective horizon designations. For example, four layers of a Bk horizon sampled by 10-cm increments are designated Bk1, Bk2, Bk3, and Bk4. If the horizon has already been subdivided because of differences in morphological features, the set of numerals that identifies the additional sampling subdivisions follows the first numeral. For example, three layers of a Bk2 horizon sampled by 10-cm increments are designated Bk21, Bk22, and Bk23. The descriptions for each of these sampling subdivisions can be the same, and a statement indicating that the horizon has been subdivided only for sampling purposes can be added.

3.6 Discontinuities

Numerals are used as prefixes to horizon designations (preceding the letters A, B, C, and R) to indicate discontinuities in mineral soils. These prefixes are distinct from the numerals that are used as suffixes denoting vertical subdivisions. A discontinuity that can be identified by a number prefix is a significant change in particle-size distribution or mineralogy that indicates a difference in the material from which the horizons have formed and/or a significant difference in age, *unless that difference in age is indicated by the suffix b.*

Symbols that identify discontinuities are used only when they can contribute substantially to an understanding of the relationships among horizons. The stratification common to soils that formed in alluvium is not designated as a discontinuity, unless particle-size distribution differs markedly from layer to layer (i.e., particle-size classes are strongly contrasting), even though genetic horizons may have formed in the contrasting layers.

Where a soil has formed entirely in one kind of material, the whole profile is understood to be material 1 and the number prefix is omitted from the symbol. Similarly, the uppermost material in a profile consisting of two or more contrasting materials is understood to be material 1, but the number is omitted. Numbering starts with the second layer of contrasting material, which is designated 2. Underlying contrasting layers are numbered consecutively. Even when the material of a layer below material 2 is similar to material 1, it is designated 3 in the sequence; the numbers indicate a change in materials, not types of material. Where two or more consecutive horizons have formed in the same kind of material, the same prefix number is applied to all the designations of horizons in that material: A-Bk1-2Bk2-2Bk3-2Ck. The suffix numbers designating vertical subdivisions of the Bk horizon continue in consecutive order across the discontinuity.

However, vertical subdivisions do not continue across lithologic discontinuities if the horizons are not consecutive or contiguous to each other. If other horizons intervene, another vertical numbering sequence begins for the lower horizons: A-C1-C2-2Bk1-2Bk2-2Ck1-2Ck2. If an R layer is present below a soil that has formed in residuum and if the material of the R layer is judged to be like the material from which the soil has developed, the Arabic number prefix is not used. The prefix is used, however, if it is thought that the R layer would produce material unlike that in the solum, e.g., A-Bk-Ck-2R or A-Bk-2R. If part of the solum has formed in residuum, the symbol R is given the appropriate prefix: A-Bk1-2Bk2-2Bk3-2Ck1-2Ck2-2R.

A buried horizon (designated by the letter b) presents special problems. It is obviously not in the same deposit as the overlying horizons. Some buried horizons, however, have formed in material that is lithologically like the overlying deposit. A prefix is not used to distinguish material of such a buried horizon. If the material in which a horizon of a buried soil has formed is lithologically unlike the overlying material, however, the discontinuity is indicated by a number prefix and the symbol for the buried horizon also is used, e.g., A-Bk1-Bk2-C-2Ab-2Bkb1-2Bkb2-2C.

3.7 Use of the Prime Symbol

If two or more horizons with *identical* numeral prefixes and letter combinations are separated by one or more horizons with a different horizon designation in a pedon, identical letter and number symbols can be used for those horizons that have the same characteristics. For example, the sequence A-Bk-Bkn-Bk-CK identifies a soil that has two Bk horizons. To emphasize this characteristic, the prime symbol (′) is added after the master-horizon symbol of the lower of the two horizons that have identical designations, e.g., A-Bk-Bkn-B′k-Ck. The prime symbol, where appropriate, is placed after the capital-letter horizon designation and before the lowercase suffix letter symbols that follow it: B′k.

The prime symbol is not used unless all letters and numeral prefixes are completely identical. The sequence A-Bk1-Bk2-2Bk1-2Bk2 is an example. It has two Bk master horizons of different lithologies; thus, the Bk horizons are not identical and the prime symbol is not needed. The prime symbol is used for soils with lithologic discontinuities when horizons have identical designations. In the rare cases where three layers have identical letter symbols, double prime symbols can be used for the lowest of these horizons: Bk″.

Vertical subdivisions of horizons or layers (numeral suffixes) are not taken into account when the prime symbol is assigned. The sequence A-Bk-Bkn-B′k1-B′k2-B′k3-C is an example.

3.8 Use of the Caret Symbol

The "caret" symbol (^) is used as a prefix to master horizon designations to indicate layers of human-transported material. This material has been moved horizontally onto a pedon from a source area outside of that pedon by directed human activity, usually with the aid of machinery.

All horizons and layers formed in human-transported material are indicated by a "caret" prefix (e.g., ^A-^C-Ak-Bkb). When they can contribute substantially to an understanding of the relationship of the horizons or layers, numeral prefixes may be used before the caret symbol to indicate the presence of discontinuities within the human-transported material or between the human-transported material and underlying layers (e.g., ^A-^C-2^C-3Bkb).

3.9 Diagnostic Surface Horizons

3.9.1 The Epipedon

The epipedon (Gr. *epi*, over, upon, and *pedon*, soil) is a horizon that forms at or near the surface and in which most of the rock structure has been destroyed. It is darkened by organic matter or shows evidence of eluviation, or both. *Rock structure* as used here and in other places in this taxonomy includes fine stratification (5 mm or less thick) in unconsolidated sediments (eolian, alluvial, or marine) and saprolite derived from consolidated rocks in which the unweathered minerals and pseudomorphs of weathered minerals retain their relative positions to each other.

Any horizon may be at the surface of a truncated soil. These horizons can be covered by a surface mantle of new soil material. If the surface mantle has rock structure, the top of the epipedon is considered the soil surface unless the mantle meets the definition of buried soils. If the soil includes a buried soil, the epipedon, if any, is at the soil surface and the epipedon of the buried soil is considered a buried epipedon and is not considered in selecting taxa unless the keys specifically indicate buried horizons. A soil with a mantle thick enough to have a buried soil has no epipedon if the soil has rock structure to the surface or has an A horizon less than 25 cm thick that is underlain by soil material with rock structure.

A recent alluvial or eolian deposit that retains fine stratifications (5 mm or less thick) or an A horizon directly underlain by such stratified material is not included in the concept of the epipedon because time has not been sufficient for soil-forming processes to erase these transient marks of deposition and for diagnostic and accessory properties to develop. An epipedon is not the same as an A horizon.

3.9.1.1 Ochric Epipedon

The ochric epipedon is the only diagnostic surface horizon recognized in the UAE. It fails to meet any of the definitions for the other epipedons identified in the USDA *Keys to Soil Taxonomy* because it is too thin or too dry, has too high a color value or chroma, contains too little organic carbon, has too high an *n* value or melanic index, or is both massive and hard or harder when dry. The ochric epipedon does not have rock structure and does not include finely stratified fresh sediments, nor can it be an Ap horizon directly overlying such deposits.

3.10 Diagnostic Subsurface Horizons

The horizons described in this section form below the surface of the soil. They may be exposed at the surface by truncation of the soil. Some of these horizons are generally regarded as B horizons, some are considered B horizons by many but not all pedologists, and others are generally regarded as parts of the A horizon. The following diagnostic subsurface horizons have been identified in the UAE.

3.10.1 Anhydritic Horizon[†]

The anhydritic diagnostic horizon (Abdelfattah and Shahid 2007; Abdelfattah et al. 2008, 2009; Shahid 2010; Shahid and Abdelfattah 2008, 2009; Shahid et al. 2004, 2007, 2009, 2010; Wilson et al. 2013) is in the final stage of approval to be incorporated as part of USDA Soil Taxonomy. The criteria are as follows:

The anhydritic horizon is an illuvial horizon in which secondary anhydrite has accumulated to a significant extent.

Fig. 3.1 Well developed anhydritic horizon

Required Characteristics
An anhydritic horizon has *all* of the following properties:

1. A thickness of 15 cm or more; *and*
2. 5 % or more (by weight) anhydrite; *and*
3. Has hue of 5Y with chroma (dry and moist) of 1 or 2, and value of 7 or 8; *and*
4. A product of thickness, in cm, multiplied by the anhydrite content (percent by weight) of 150 or more. Thus, a horizon 30 cm thick that is 5 % anhydrite qualifies as an anhydritic horizon, *and*
5. Anhydrite should be the predominant mineral, with gypsum either absent or occurring only in minor amounts (Fig. 3.1).

3.10.2 Calcic Horizon

The calcic horizon is an illuvial horizon in which secondary calcium carbonate or other carbonates have accumulated to a significant extent.

Required Characteristics
The calcic horizon:

1. Is 15 cm or more thick; *and*
2. Has *one or more* of the following:
   or
   (a) 15 % or more (by weight) $CaCO_3$ equivalent and 5% or more (by volume) identifiable secondary carbonates; *or*
   (b) 5 % or more (by weight) calcium carbonate equivalent and:
   (1) Has less than 18 % clay in the fine-earth fraction; *and*
   (2) Meets the criteria for a sandy, sandy-skeletal, coarse-loamy, or loamy-skeletal particle-size class; *and*
3. Is not cemented or indurated in any part by carbonates, with or without other cementing agents, or is cemented in some part and the cemented part satisfies *one* of the following:
   (a) It is characterized by so much lateral discontinuity that roots can penetrate through noncemented zones or along vertical fractures with a horizontal spacing of less than 10 cm; *or*
   (b) The cemented layer is less than 1 cm thick and consists of a laminar cap underlain by a lithic or paralithic contact; *or*
   (c) The cemented layer is less than 10 cm thick.

3.10.3 Cambic Horizon

A cambic horizon is the result of physical alterations, chemical transformations, or removals or of a combination of two or more of these processes.

Required Characteristics
The cambic horizon is an altered horizon 15 cm or more thick. In addition, the cambic horizon must meet all of the following:

1. Has a texture class of very fine sand, loamy very fine sand, or finer; *and*
2. Shows evidence of alteration in the following forms:
   (a) Has soil structure, *or*
   (b) The absence of rock structure, including fine stratifications (5 mm or less thick), in more than one-half of the volume and one or more of the following properties:
   (1) Higher chroma, higher value, redder hue, or higher clay content than the underlying horizon or an overlying horizon; *or*

(2) Evidence of the removal of carbonates or gypsum; *and*

3. Has properties that do not meet the requirements for a calcic, gypsic, petrocalcic, or petrogypsic; *and*
4. Is not part of an Ap horizon.

3.10.4 Gypsic Horizon

The gypsic horizon is a horizon in which gypsum has accumulated or been transformed to a significant extent. It typically occurs as a subsurface horizon, but it may occur at the surface in some soils.

Required Characteristics
A gypsic horizon meets all of the following requirements:

1. Is 15 cm or more thick; *and*
2. Is not cemented by gypsum, with or without other cementing agents; is cemented and the cemented parts are less than 5 mm thick; or is cemented but, because of lateral discontinuity, roots can penetrate along vertical fractures with a horizontal spacing of less than 10 cm; *and*
3. Is 5 % or more (by weight) gypsum and has 1 % or more (by volume) visible secondary gypsum that has either accumulated or been transformed; *and*
4. Has a product of thickness, in cm, multiplied by the gypsum content (percent by weight) of 150 or more. Thus, a horizon 30 cm thick that is 5 % gypsum qualifies as a gypsic horizon if it is 1 % or more (by volume) visible gypsum and any cementation is as described in 2 above (Fig. 3.2).

3.10.5 Petrocalcic Horizon

The petrocalcic horizon is an illuvial horizon in which secondary calcium carbonate or other carbonates have accumulated to the extent that the horizon is cemented or indurated.

Required Characteristics
A petrocalcic horizon must meet the following requirements:

1. The horizon is cemented or indurated by carbonates, with or without silica or other cementing agents; *and*

Fig. 3.2 Example of gypsic horizon

2. Because of lateral continuity, roots can penetrate only along vertical fractures with a horizontal spacing of 10 cm or more; *and*
3. The horizon has a thickness of:
   (a) 10 cm or more; *or*
   (b) 1 cm or more if it consists of a laminar cap directly underlain by bedrock (Fig. 3.3).

3.10.6 Petrogypsic Horizon

The petrogypsic horizon is a horizon in which visible secondary gypsum has accumulated or has been transformed. The horizon is cemented (i.e., extremely weakly through indurated cementation classes), and the cementation is both laterally continuous and root limiting, even when the soil is moist. The horizon typically occurs as a subsurface horizon, but it may occur at the surface in some soils.

Fig. 3.3 Example of petrocalcic horizon

Required Characteristics

A petrogypsic horizon meets *all* of the following requirements:

1. Is cemented or indurated by gypsum, with or without other cementing agents (Fig. 3.4); *and*
2. Because of lateral continuity, can be penetrated by roots only along vertical fractures with a horizontal spacing of 10 cm or more; *and*
3. Is 5 mm or more thick; *and*
4. Is 40 % or more (by weight) gypsum.

3.10.7 Salic Horizon

A salic horizon is a horizon of accumulation of salts that are more soluble than gypsum in cold water.

Fig. 3.4 Example of strongly cemented petrogypsic horizon

Required Characteristics
A salic horizon is 15 cm or more thick and has, for 90 consecutive days or more in normal years:

1. An electrical conductivity (EC) equal to or greater than 30 dS/m in the water extracted from a saturated paste; *and*
2. A product of the EC, in dS m^{-1}, and thickness, in cm, equal to 900 or more.

3.11 Horizons and Characteristics-Diagnostics for Mineral Soils

Following are descriptions of the horizons and characteristics that are diagnostic for mineral soils of the Unites Arab Emirates.

3.11.1 Aquic Conditions

Soils with aquic (L. *aqua*, water) conditions are those that currently undergo continuous or periodic saturation and reduction. The presence of these conditions is indicated by redoximorphic features and can be verified by measuring saturation and reduction, except in artificially drained soils. Artificial drainage is defined here as the removal of free water from soils having aquic conditions by surface mounding, ditches, or subsurface tiles or the prevention of surface or ground water from reaching the soils by dams, levees, surface pumps, or other means. In these soils water table levels and/or their duration are changed significantly in connection with specific types of land use. Upon removal of the drainage practices, aquic conditions would return. In the keys, artificially drained soils are included with soils that have aquic conditions.

Fig. 3.5 Profile showing lithic contact at 45 cm depth (AD113)

3.11.2 Lithic Contact

A lithic contact is the boundary between soil and a coherent underlying material (Fig. 3.5). The underlying material must be virtually continuous within the limits of a pedon. Cracks that can be penetrated by roots are few, and their horizontal spacing is 10 cm or more. The underlying material must be sufficiently coherent when moist to make hand-digging with a spade impractical, although the material may be chipped or scraped with a spade. The material below a lithic contact must be in a strongly cemented or more cemented rupture-resistance class. Commonly, the material is indurated. The underlying material considered here does not include diagnostic soil horizons, such as a duripan or a petrocalcic horizon. A lithic contact is diagnostic at the subgroup level if it is within 50 cm of the surface for the soils which occur in the UAE.

3.11.3 Paralithic Contact

A paralithic (lithic-like) contact is a contact between soil and paralithic materials (defined below) where the paralithic materials have no cracks or the spacing of cracks that roots can enter is 10 cm or more.

3.11.4 Paralithic Materials

Paralithic materials are relatively unaltered materials (do not meet the requirements for any other named diagnostic horizons or any other diagnostic soil characteristic) that have an extremely weakly cemented to moderately cemented rupture-resistance class. Cementation, bulk density, and the organization are such that roots cannot enter, except in cracks. Paralithic materials have, at their upper boundary, a paralithic contact if they have no cracks or if the spacing of cracks that roots can enter is 10 cm or more. Commonly, these materials are partially weathered bedrock or weakly consolidated bedrock, such as sandstone, siltstone, or shale. Paralithic materials can be used to differentiate soil series if the materials are within the series control section. Fragments of paralithic materials 2.0 mm or more in diameter are referred to as pararock fragments.

3.11.5 Soil Moisture Regimes

The term "soil moisture regime" refers to the presence or absence either of ground water or of water held at a tension of less than 1,500 kPa in the soil or in specific horizons during periods of the year. Water held at a tension of 1,500 kPa or more is not available to keep most mesophytic plants alive. The availability of water is also affected by dissolved salts. If a soil is saturated with water that is too salty to be available to most plants, it is considered salty rather than dry. Consequently, a horizon is considered dry when the moisture tension is 1,500 kPa or more and is considered moist if water is held at a tension of less than 1,500 kPa but more than zero. A soil may be continuously moist in some or all horizons either throughout the year or for some part of the year. It may be either moist in winter and dry in summer or the reverse. In the Northern Hemisphere, summer refers to June, July, and August and winter refers to December, January, and February.

Normal Years

A normal year is defined as a year that has:

1. Annual precipitation that is plus or minus one standard deviation of the long-term (30 years or more) mean annual precipitation; *and*
2. Mean monthly precipitation that is plus or minus one standard deviation of the long-term monthly precipitation for 8 of the 12 months.

For the most part, normal years can be calculated from the mean annual precipitation; however, when catastrophic events occur during a year, the standard deviations of the monthly means should also be calculated. The term "normal years" replaces the terms "most years" and "6 out of 10 years," which were used in the 1975 edition of *Soil Taxonomy* (USDA, SCS 1975). When precipitation data are evaluated to determine if the criterion for the presence of aquic conditions, or number of days that the moisture control section is moist, or number of days that some part of the soil is saturated has been met, it is permissible to include data from periods with below normal rainfall. Similarly, when precipitation data are evaluated to determine if the criterion for the number of days that the moisture control section is dry has been met, it is permissible to include data from periods with above normal rainfall. It is assumed that if the criteria are met during these periods, they will also be met during normal years.

3.11.5.1 Classes of Soil Moisture Regimes

The soil moisture regimes are defined in terms of the level of ground water and in terms of the seasonal presence or absence of water held at a tension of less than 1,500 kPa in the moisture control section. It is assumed in the definitions that the soil supports whatever vegetation it is capable of supporting, i.e., crops, grass, or native vegetation, and that the amount of stored moisture is not being increased by irrigation or fallowing. These cultural practices affect the soil moisture conditions as long as they are continued.

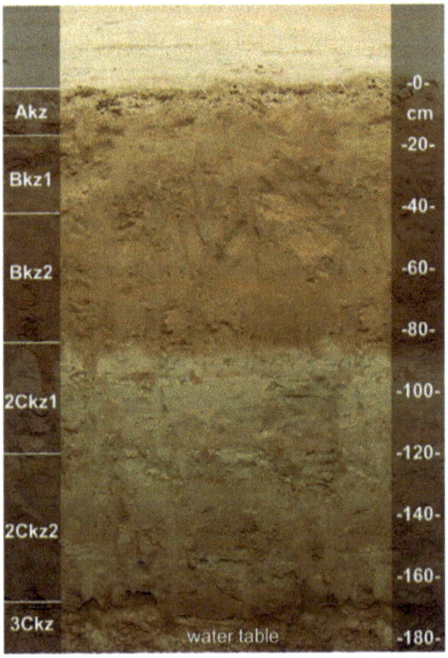

Fig. 3.6 Profile showing aquic moisture regime (AD147)

Aquic soil moisture regime.—The aquic (L. *aqua*, water) soil moisture regime is a reducing regime in a soil that is virtually free of dissolved oxygen because it is saturated by water (Fig. 3.6). Some soils are saturated with water at times while dissolved oxygen is present, either because the water is moving or because the environment is unfavorable for micro-organisms (e.g., if the temperature is less than 1°C); such a regime is not considered aquic.

It is not known how long a soil must be saturated before it is said to have an aquic soil moisture regime, but the duration must be at least a few days, because it is implicit in the concept that dissolved oxygen is virtually absent. Because dissolved oxygen is removed from ground water by respiration of microorganisms, roots, and soil fauna, it is also implicit in the concept that the soil temperature is above biologic zero for some time while the soil is saturated. Biologic zero is defined as 5°C in this taxonomy.

Aridic and torric (L. *aridus*, dry, and L. *torridus*, hot and dry) soil moisture regimes.—These terms are used for the same moisture regime but in different categories of the taxonomy. In the aridic (torric) soil moisture regime, the moisture control section is, in normal years:

1. Dry in all parts for more than half of the cumulative days per year when the soil temperature at a depth of 50 cm below the soil surface is above 5°C; *and*
2. Moist in some or all parts for less than 90 consecutive days when the soil temperature at a depth of 50 cm below the soil surface is above 8°C.

Soils that have an aridic (torric) soil moisture regime normally occur in areas of arid climates. A few are in areas of semiarid climates and either have physical properties that keep them dry, such as a crusty surface that virtually precludes the infiltration of water, or are on steep slopes where runoff is high. There is little or no leaching in this soil moisture regime, and soluble salts accumulate in the soils if there is a source.

3.11.6 Soil Temperature Regime

Following is a description of the soil temperature regime identified in the UAE and used in defining the hyperthermic soil temperature class which is used at the family categorical level of soil taxonomy.

Hyperthermic.—The mean annual soil temperature is 22°C or higher, and the difference between mean summer and mean winter soil temperatures is 6°C or more either at a depth of 50 cm below the soil surface or at a densic, lithic, or paralithic contact, whichever is shallower.

References

Abdelfattah, M. A., & Shahid, S. A. (2007). A comparative characterization and classification of soils in Abu Dhabi coastal area in relation to arid and semi-arid conditions using USDA and FAO soil classification systems. *Arid Land Research and Management 21*, 245–271.

Taylor and Francis Publishers. Full paper is available online at: http://www.informaworld.com/smpp/content~content=a779704000~db=all~order=page. Accessed 3 May 2011.

Abdelfattah, M. A., Shahid, S. A., & Wilson, M. A. (2008). *Characterization of anhydrite soils discovered in the coastal Sabkha of United Arab Emirates*. Joint Annual Meeting of the GSA-ASA-CSSA-SSSA-GCAGS-GCSSEPM, 5–9 Oct 2008, Houston. Abstracts are available online at: http://a-c-s.confex.com/crops/2008am/webprogram/Paper50710.html. Accessed 3 May 2011.

Abdelfattah, M. A., Shahid, S. A., & Wilson, M. A. (2009). *Gypseous and anhydritic soils-proposed changes in the USDA soil taxonomy*. Proceedings of the 10th annual research conference, UAE University (Vol. I, pp. 167–176). Book of abstracts is available online at: http://sra.uaeu.ac.ae/Conference_10/Proceedings/Book_of_Abstracts_II.pdf. Accessed 3 May 2011.

Shahid, S. A. (2010). Breakthrough in soil taxonomy. New information for the international soil science community. Technical flyer published by International Center for Biosaline Agriculture, Dubai.

Shahid, S. A., & Abdelfattah, M. A. (2008). Soils of Abu Dhabi Emirate. In R. J. Perry (Ed.), *Terrestrial environment of Abu Dhabi Emirate* (pp. 71–92). Environment Agency–Abu Dhabi, UAE.

Shahid, S. A., & Abdelfattah, M. A. (2009). Soil information breakthrough in Abu Dhabi—new soil information for the international soil science community. *Biosalinity News, 10*(1), 6–7.

Shahid, S. A., Abdelfattah, M. A., Arshad, K. R., Muhairbi, M., Al Othman, Y. R., Al Haji, A., Mansouri, M., Al Ahmad, S., Siddiqui, A., Azimullah, K. M., Sekaly, A. L., & Al Qubaisi, B. S. (2004). *Soil survey for the coastline of Abu Dhabi Emirate. Two volumes (Vol. 1: Reconnaissance Survey; Vol. 2: Soil Maps)*. Abu Dhabi: Unpublished report of the Environmental Research and Wildlife Development Agency.

Shahid, S. A., Abdelfattah, M. A., & Wilson, M. A. (2007). A unique anhydrite soil in the coastal sabkha of Abu Dhabi Emirate. *Soil Survey Horizons, 48*, 75–79.

Shahid, S. A., Abdelfattah, M. A., Al-Saiy, A. K., & Mufti, K. A. (2009). Commercial value assessment of newly discovered anhydrite soil in the coastal land of Abu Dhabi Emirate. *European Journal of Scientific Research, 29*(1), 36–46.

Shahid, S. A., Taha, F. K., & Abdelfattah, M. A. 2010. Synthesis report (unpublished). International conference on soil classification and reclamation of degraded agricultural lands in arid environments (p. 13). Fairmont Abu Dhabi, 17–19 May 2010

Soil Survey Staff. (2010). *Keys to soil taxonomy* (11th ed.). Washington, DC: U.S. Department of Agriculture, Natural Resources Conservation Service. U.S. Government Printing Office.

USDA, SCS. (1975). *Soil taxonomy: A basic system of soil classification for making and interpreting soil surveys* (U.S. Department of Agriculture Handbook, No. 436). Washington, DC: U.S. Government Printing Office.

Wilson, M. A., Shahid, S. A., Abdelfattah, M. A., Kelley, J. A., & Thomas, J. E. (2013). Anhydrite formation on the coastal sabkha of Abu Dhabi, United Arab Emirates. Chapter 8. In S. A. Shahid, F. K. Taha, & M. A. Abdelfattah (Eds.), *Developments in soil classification, land use planning and policy implications* (pp. 123–140). Springer Dordrecht Heidelberg New York London.

Field kit used to establish soil taxa

Greening the desert to combat desertification in UAE

Family and Series Differentiae

4

Abstract

This chapter describes families and series identified in the UAE. The families and series are established using various criteria for mineral soils. The particle-size classes and their substitutes, mineralogy classes, cation-exchange activity classes, soil temperature classes, soil depth classes of the control section of soil taxa are used. Particle-size classes (Fragmental, Gypseous-skeletal, Coarse-gypseous, Fine gypseous, Sandy-skeletal, Loamy-skeletal, Sandy, loamy, Coarse-loamy, Fine-loamy, Coarse-silty, Fine-silty, fine), Mineralogy classes (hypergypsic, anhydritic, gypsic, carbonatic and mixed), Cation-exchange activity classes (superactive, active, semiactive, subactive), Soil temperature class (hyperthermic) and Soil depth class (shallow), Key to phases of soil classes (lithic, salic, petrogypsic, petrocalcic, anhydritic, gypsic, calcic, salidic, sodic, aquic, shelly) identified in the UAE are described.

Keywords

Soil family • Particle size classes • Mineralogy classes • Cation exchange activity • Soil depth classes

4.1 Introduction

The following differentiae are used to distinguish families of mineral soils. The class names of these components are used to form the family name. The components are listed and defined in the same sequence in which the components appear in the family names. When written, family classes are separated by a comma (Typic Torripsamments, carbonatic, hyperthermic, *salic*). It should be noted that not all families are provided in the name for each subgroup. For example, since

S.A. Shahid et al., *United Arab Emirates Keys to Soil Taxonomy*,
DOI 10.1007/978-94-007-7420-9_4,

Torripsamments (by definition) are sandy, it would be redundant to list the particle size class in the family name. If more than one phase is needed, they are separated by a hyphen (Typic Torripsamments, carbonatic, hyperthermic, lithic-salidic).

Particle-size classes and their substitutes
Mineralogy classes
Cation-exchange activity classes
Calcareous and reaction classes
Soil temperature classes
Soil depth classes
Phases of soil taxa

4.2 Definition of Particle-Size Classes

The term particle-size class is used to characterize the grain-size composition of the whole soil, including both the fine earth and the rock and pararock fragments up to the size of a pedon, but it excludes organic matter and salts more soluble than gypsum. The particle-size (texture) classes are based on percentages, by weight, in the fraction less than 2.0 mm in diameter.

4.2.1 Control Section for Particle-Size Classes

The particle-size and substitute class names listed below are applied to certain horizons, or to the soil materials within specific depth limits, that have been designated as the control section for particle-size classes and their substitutes. The lower boundary of the control section may be at a specified depth (in centimeters) below the mineral soil surface, or it may coincide with the upper boundary of a root-limiting layer (defined below).

4.2.1.1 Root-Limiting Layers

Unless otherwise indicated, the following are considered root-limiting layers in the UAE: a petrocalcic horizon, a petrogypsic horizon, a lithic contact, or a paralithic contact.

4.2.1.2 Key to the Control Section for Particle-Size Classes

The following list of particle-size control sections for the soils of the UAE is arranged as a key to allow the reader to make the correct classification by starting at the beginning and eliminating one by one all classes that include criteria that do not fit the soil in question. The soil belongs to the first class for which it meets all of the criteria listed.

(A) For mineral soils that have a root-limiting layer (listed above) within 36 cm of the mineral soil surface, from the mineral soil surface to the root-limiting layer; *or*

(B) Between the lower boundary of an Ap horizon or surface layer or a depth of 25 cm below the mineral soil surface, whichever is deeper, and the shallower of the following: (a) a depth of 100 cm below the mineral soil surface or (b) a root-limiting layer.

4.2.1.3 Key to the Particle-Size Classes

This key to UAE soils is arranged to allow the reader to make the correct particle-size distribution classification by going through the key systematically, starting at the beginning and eliminating one by one all classes that include criteria that do not fit the soil or layer in question. These classes have been or are expected to be identified in UAE soils.

Throughout the control section, meet one of the following sets of particle-size class criteria:

1. Mineral soils that have a fine-earth component of less than 10 % (including associated medium and finer pores) of the total volume (Fig. 4.1). **Fragmental**

   or

2. Have, in the fraction less than 20 mm in diameter, 40 % or more (by weight) gypsum *and* one of the following:

   (a) A total of 35 % or more (by volume) rock fragments. **Gypseous-skeletal**

   or

   (b) Less than 35 % (by volume) rock fragments and 50 % or more (by weight) particles with diameter of 0.1–2 mm. **Coarse-gypseous**

   or

Fig. 4.1 Profile showing fragmental particle size class (NE002)

(c) Less than 35 % (by volume) rock fragments.

Fine-gypseous

or

3. Have 35 % or more (by volume) rock fragments *and* a fine-earth fraction with a texture class of sand or loamy sand, including less than 50 % (by weight) very fine sand (Fig. 4.2).

Sandy-skeletal

or

4. Have 35 % or more (by volume) rock fragments and less than 35 % (by weight) clay.

Loamy-skeletal

or

5. Have a texture class of sand or loamy sand, including less than 50 % (by weight) very fine sand particles in the fine-earth fraction (Fig. 4.3).

Sandy

or

6. Have a texture class of loamy very fine sand, very fine sand, or finer, including less than 35 % (by weight) clay in the fine-earth fraction, and are in a shallow family or in a Lithic subgroup.

Loamy

or

7. Have, in the fraction less than 75 mm in diameter, 15 % or more (by weight) particles with diameters of 0.1–75 mm (fine sand or coarser, including gravel) *and*, in the fine-earth fraction, less than 18 % (by weight) clay (Fig. 4.4).

Coarse-loamy

or

Fig. 4.2 Profile showing sandy-skeletal particle size class (AD105)

Fig. 4.3 Profile showing sandy particle size class (AD158)

8. Have, in the fraction less than 75 mm in diameter, 15 % or more (by weight) particles with diameters of 0.1–75 mm (fine sand or coarser, including gravel) *and*, in the fine-earth fraction, 18–35 % (by weight) clay.

Fine-loamy

or

9. Have, in the fraction less than 75 mm in diameter, less than 15 % (by weight) particles with diameters of 0.1–75 mm (fine sand or coarser, including gravel) *and*, in the fine-earth fraction, less than 18 % (by weight) clay (Fig. 4.5).

Coarse-silty

10. Have in the fraction less than 75 mm in diameter, less than 15 % (by weight) particles with diameters of 0.1–75 mm (fine sand or coarser, including gravel) *and*, in the fine-earth fraction, 18–35 % (by weight) clay.

Fine-silty

or

11. Have (by weighted average) less than 60 % (by weight) clay in the fine-earth fraction.

Fine

4.2.2 Strongly Contrasting Particle-Size Classes

The purpose of strongly contrasting particle-size classes is to identify changes in pore-size distribution or composition that are not identified in

Fig. 4.4 Profile showing coarse-loamy particle size class (AD132)

Fig. 4.5 Profile showing coarse-silty particle size class (NE019)

higher soil categories and that seriously affect the movement and retention of water and/or nutrients.

The particle-size or substitute classes listed below are considered strongly contrasting if both parts are 12.5 cm or more thick (including the thickness of these parts not entirely within the particle-size control section; however, substitute class names are used only if the soil materials to which they apply extend 10 cm or more into the upper part of the particle-size control section) and if the transition zone between the two parts of the particle-size control section is less than 12.5 cm thick.

Some classes, such as sandy and sandy-skeletal, have been combined in the following list. In those cases the combined name is used as the family class if part of the control section meets the criteria for either class.

To date three strongly contrasting particle-size classes have been identified in the United Arab Emirates. The following classes are not presented in a key format.

1. Fine-gypseous over loamy: Um Al Quwain Series
2. Coarse-loamy over sandy or sandy-skeletal (if the coarse-loamy material contains less than 50 %, by weight, fine sand or coarser sand): Wahala Series
3. Sandy over loamy: As Sihebi Series

4.3 Mineralogy Classes

The mineralogy of soils is known to be useful in making predictions about soil behavior and responses to management. Some mineralogy classes occur or are important only in certain taxa or particle-size classes, and others are important in all particle-size classes.

4.3.1 Control Section for Mineralogy Classes

The control section for mineralogy classes is the same as that defined for the particle-size classes and their substitutes.

4.3.2 Key to Mineralogy Classes

This key, like other keys in this taxonomy, is designed in such a way that the reader makes the correct classification by going through the key systematically, starting at the beginning and eliminating one by one any classes that include criteria that do not fit the soil in question. The soil belongs to the first class for which it meets all of the required criteria. The user should first check the criteria listed in the first class, if the soil in question does not meet the criteria listed there, proceed on to the next class, until the soil meets the criteria listed. All criteria are based on a weighted average.

For soils with strongly contrasting particle-size classes, mineralogy classes are used for both of the named parts of particle-size classes or substitute classes, unless they are the same. The same mineralogy class name cannot be used for both parts of the control section (e.g., "mixed over mixed"). Example of a soil that requires assignment of two different mineralogy classes is a fine-gypseous over loamy, hypergypsic over carbonatic, hyperthermic Gypsic Aquisalid.

1. Soil layers or horizons, in the mineralogy control section, that have a substitute class that replaces the particle-size class, other than fragmental, and that have 40 % or more (by weight) gypsum either in the fine-earth fraction or in the fraction less than 20 mm in size, whichever has a higher percentage of gypsum.

Hypergypsic

or

2. Soil layers or horizons that have any particle-size class and 15 % or more (by weight) anhydrite, either in the fine-earth fraction or in the fraction less than 20 mm in size, whichever has a higher percentage of anhydrite (Fig. 4.6).

Anhydritic

or

3. Soil layers or horizons, in the mineralogy control section, that have any particle-size class and 15 % or more (by weight) gypsum, either in the fine-earth fraction (less than 2 mm)

Fig. 4.6 Soil profile showing anhydritic mineralogy class (AD125)

or in the fraction less than 20 mm in size, whichever has a higher percentage of gypsum (Fig. 4.7).

Gypsic

or

4. Soil layers or horizons that have any particle-size class and more than 40 % (by weight) carbonates (expressed as $CaCO_3$) plus gypsum plus anhydrite, either in the fine-earth fraction or in the fraction less than 20 mm in size, whichever has a higher percentage of carbonates plus gypsum plus anhydrite (Fig. 4.8).

Carbonatic

or

5. Soils with mineralogy classes other than hypergypsic, anhydritic, gypsic, carbonatic, or siliceous.

Mixed

4.4 Cation-Exchange Activity Classes

The cation-exchange activity classes help in making interpretations of mineral assemblages and of the nutrient holding capacity of soils in mixed mineralogy classes of loamy-skeletal, loamy, coarse-loamy, fine-loamy, fine-silty, and fine particle-size classes.

Cation-exchange activity classes are not assigned to Psamments, "psamm" great groups of Entisols, or other soils with sandy or sandy-skeletal particle-size classes or the fragmental substitute

Fig. 4.7 Soil profile showing gypsic mineralogy class (AD111)

Fig. 4.8 Soil profile showing carbonatic mineralogy class (AD156)

class because the low clay content causes cation-exchange activity classes to be less useful and less reliable.

The cation-exchange capacity is determined by NH_4OAc at pH 7 on the fine-earth fraction. The CEC of the organic matter, sand, silt, and clay is included in the determination.

The criteria for the classes use ratios of CEC to the percent, by weight, of silicate clay, both by weighted average in the control section. In the following classes "clay" excludes clay-size carbonates. If the ratio of percent water retained at 1,500 kPa tension to the percentage of measured clay is 0.25 or less or 0.6 or more in half or more of the particle-size control section (or part in contrasting families), then the percentage of clay is estimated by the following formula: Clay % = 2.5(% water retained at 1,500 kPa tension—% organic carbon).

4.4.1 Control Section for Cation-Exchange Activity Classes

The control section for cation-exchange activity classes is the same as that used to determine the particle-size and mineralogy classes. For soils with strongly contrasting particle-size classes, where both named parts of the control section use a cation exchange activity class, the class associated with the particle size class that has the most clay is named.

4.4.2 Key to Cation-Exchange Activity Classes

A. Soils that are not Psamments, and that are not in a sandy or sandy-skeletal particle-size class or any substitute for a particle-size class (including fragmental), and that have:

1. A mixed mineralogy class; and
2. A ratio of cation-exchange capacity (by IN NH_4OAc pH 7) to percent clay (by weight) of:
   (a) 0.60 or more.

Superactive

   (b) 0.40 to less than 0.60

Active

(c) 0.24 to less than 0.40

Semiactive

(d) Less than 0.24

Subactive

4.5 Soil Temperature Classes

Soil temperature class, as named and defined here, is used as part of the family name. The Celsius (centigrade) scale is the standard. It is assumed that the temperature is that of a soil that is not being irrigated.

4.5.1 Control Section for Soil Temperature

The control section for soil temperature is either at a depth of 50 cm below the soil surface or at the upper boundary of a root-limiting layer, whichever is shallower. The soil temperature class is defined in terms of the mean annual soil temperature and the difference between mean summer and mean winter temperatures.

The only soil temperature class identified in the UAE is the **hyperthermic** class. It is defined for soils that have a difference in soil temperature of 6°C or more between mean summer (June, July, and August) and mean winter (December, January, and February) and a mean annual soil temperature of 22°C (72°F) or higher.

Hyperthermic

4.6 Soil Depth Classes

Soil depth class is used in families of United Arab Emirates mineral soils that have a root-limiting layer at a specified depth from the mineral soil surface. The root-limiting layers included in soil depth classes are petrocalcic or petrogypsic horizons and paralithic contact. Shallow soils that meet the criteria for «lithic contact» using the USDA-*Keys to Soil Taxonomy* (11th and previous editions, Soil Survey Staff 2010) have now been assigned to the appropriate "Lithic" subgroups in the *UAE Keys to Soil Taxonomy*.

4.6.1 Key to Soil Depth Classes

(A) Mineral soils that are less than 50 cm deep (from the soil surface) to a root-limiting layer (petrogypsic or petrocalcic horizon, or a paralithic contact) excluding soils that are in a Lithic subgroup (Fig. 4.9).

Shallow

(B) All other mineral soils: No soil depth class used.

Fig. 4.9 Example showing shallow phase (AD124)

4.7 Phases of Soil Taxa

Phases of soil taxa have been developed for those mineral soils that have soil properties or characteristics that occur at a deeper depth than currently identified for an established taxonomic subgroup or soil properties that effect interpretations not currently recognized at the subgroup level. The phases which have been identified in the UAE include: anhydritic, aquic, calcic, gypsic, lithic, petrocalcic, petrogypsic, salic, salidic, shelly, and sodic.

Phases are added to the family classification separated by a comma (Typic Torripsamments, carbonatic, hyperthermic, salic). If more than one contrasting phase is needed, the phases are separated by a hyphen and listed in the same order as the Key (Typic Torripsamments, carbonatic, hyperthermic, lithic-salidic). Rarely are more than two phases used.

Phases are not repeated when the soil feature is recognized at the subgroup level even though it also occurs at a lower depth (e.g., Salidic Haplocalcids, sandy, carbonatic, hyperthermic, salidic). However, they are used if different than the subgroup name (e.g., Salidic Haplocalcids, sandy, carbonatic, hyperthermic, salic).

4.7.1 Key to Phases

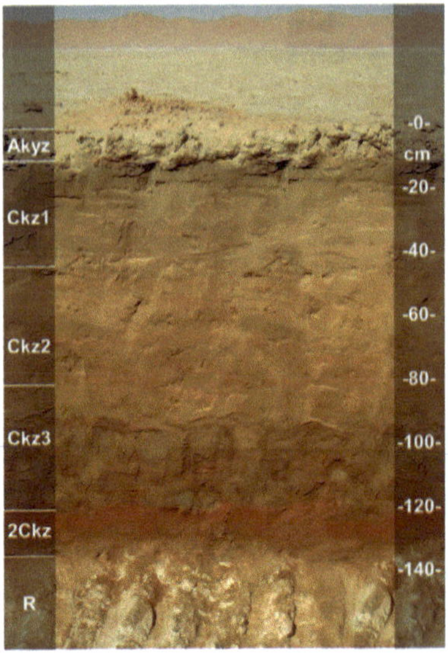

Fig. 4.10 Example showing lithic phase (AD148)

(A) Mineral soils that have a lithic contact at a depth of more than 50–200 cm from the soil surface (Fig. 4.10).

Lithic

or

(B) Mineral soils that have, at a depth of more than 100–200 cm from the soil surface, one or more of the following:

1. A salic horizon.

Salic

or

2. A petrogypsic horizon.

Petrogypsic

or

3. A petrocalcic horizon.

Petrocalcic

or

4. An anhydritic horizon.

Anhydritic[1]

or

5. A gypsic horizon.

Gypsic

or

6. A calcic horizon.

Calcic

or

7. An ECe of more than 8 to less than 30 dS m^{-1} in a layer 10 cm or more thick.

Salidic

or

[1] For the *Soil Survey of the Abu Dhabi Emirate—Extensive Survey*, the "anhydritic" phase represented those soils that contained large amounts of anhydrite in place of gypsum within a depth of 100 cm of the soil surface. These soils were classified in a "Gypsic" subgroup because (at the time of publication) the anhydritic subgroup had not been established. Currently, the anhydritic phase is used for those soils with an anhydritic horizon at a depth of more than 100–200 cm from the soil surface.

8. An exchangeable sodium percentage of 15 or more (or an SAR of 13 or more) during at least 1 month in normal years and an ECe of 8 dS m^{-1} or less in any layer with a thickness of 10 cm or more:

Sodic

or

9. Are saturated with water in one or more layers for 1 month or more in normal years.

Aquic

or

(C) Mineral soils that have, within a depth of 200 cm from the soil surface, a horizon 15 cm or more thick with 35 % or more (by volume) broken or unbroken sea shell fragments.

Shelly

Reference

Soil Survey Staff. (2010). *Keys to soil taxonomy* (11th ed.). Washington, DC: U.S. Department of Agriculture, Natural Resources Conservation Service. U.S. Government Printing Office.

Barchan dunes (Entisols) cover large part of the UAE

Identification of the Taxonomic Class of a Soil

5

Abstract

In this chapter, key to soil taxonomy hierarchy of UAE, from order-suborder-great group-subgroups are described. Within the soil order Aridisols, the soil taxa have been described as suborders, great groups and subgroups, such as, Suborders (Salids, Gypsids, Calcids, Cambids), Great Groups (Petrocalcids, Haplocalcids, Petrogypsids, Haplocambids, Calcigypsids, Haplogypsids, Aquisalids, Haplosalids), and Subgroups (Salidic Lithic Haplocalcids, Lithic Haplocalcids, Salidic Haplocalcids, Sodic Haplocalcids, Typic Haplocalcids, Calcic Petrocalcids, Typic Petrocalcids, Sodic Haplocambids, Fluventic Haplocambids, Typic Haplocambids, Lithic Calcigypsids, Salidic Calcigypsids, Typic Calcigypsids, Salidic Lithic Haplogypsids, Lithic Haplogypsids, Salidic Leptic Haplogypsids, Leptic Haplogypsids, Salidic Haplogypsids, Typic Haplogypsids, Salidic Lithic Petrogypsids, Lithic Petrogypsids, Salidic Calcic Petrogypsids, Calcic Petrogypsids, Salidic Petrogypsids, Typic Petrogypsids, Gypsic Lithic Aquisalids, Lithic Aquisalids, Petrogypsic Aquisalids, Anhydritic Aquisalids, Gypsic Aquisalids, Typic Aquisalids, Petrogypsic Lithic Haplosalids, Gypsic Lithic Haplosalids, Calcic Lithic Haplosalids, Lithic Haplosalids, Petrogypsic Haplosalids, Anhydritic Haplosalids, Leptic Gypsic Haplosalids, Gypsic Haplosalids, Calcic Haplosalids, and Typic Haplosalids). Within the order Entisols, the soil taxa have been described as Suborders (Aquents, Psamments, Orthents), Great Groups (Psammaquents, Torriorthents, Torripsamments), Subgroups (Salidic Psammaquents, Salidic Lithic Torriorthents, Lithic Torriorthents, Salidic Torriorthents, Sodic Torriorthents, Typic Torriorthents, Salidic Lithic Torripsamments, Lithic Torripsamments, Oxyaquic Torripsamments, Salidic Torripsamments, Sodic Torripsamments, and Typic Torripsamments).

S.A. Shahid et al., *United Arab Emirates Keys to Soil Taxonomy*,
DOI 10.1007/978-94-007-7420-9_5, © Springer Science+Business Media Dordrecht (Outside the USA) 2014

Keywords
Soil order • Suborder • Great group • Sub group • Soil taxa

5.1 Introduction

In the United Arab Emirates, Aridisols and Entisols soil orders have been recognized (Soil Survey Staff 2010). Most of the natural desert landscapes are not in use (Fig. 5.1), in places agriculture has been established using smart irrigation systems (Fig. 5.2).

Fig. 5.1 Natural desert landscape with Aridisols (foreground) and Entisols (background)

5.2 Key to Soil Orders

The soils that:

1. Have:
   (a) An aridic soil moisture regime; *and*
   (b) An ochric epipedon; *and*
   (c) One or more of the following within 100 cm of the soil surface: a cambic horizon with a lower depth of 25 cm or more; an anhydritic horizon; a calcic, gypsic, petrocalcic, petrogypsic, or salic horizon; *or*
2. Have a salic horizon; *and*
   (a) Saturation with water in one or more layers within 100 cm of the soil surface for 1 month or more during a normal year; *and*

Fig. 5.2 Typical desert landscape with proper land management and irrigated agriculture

(b) A moisture control section that is dry in some or all parts at some time during normal years.

Aridisols, p. 53

or

Other soils that do not classify as Aridisols.

Entisols, p. 70

5.3 Key to the Suborders of Aridisols

Aridisols that have a salic horizon within 100 cm of the soil surface.

Salids, p. 63

Other Aridisols that have a gypsic or petrogypsic horizon within 100 cm of the soil surface and do not have a petrocalcic horizon overlying these horizons.

Gypsids, p. 57

Other Aridisols that have a calcic or petrocalcic horizon within 100 cm of the soil surface.

Calcids, p. 54

Other Aridisols that have a cambic horizon within 100 cm of the soil surface.

Cambids, p. 56

5.3.1 Calcids

Key to Great Groups
Calcids that have a petrocalcic horizon within 100 cm of the soil surface.

Petrocalcids, p. 55

Other soils that are the most extensive of the Calcids and do not have a petrocalcic horizon with its boundary within 100 cm of the soil surface.

Haplocalcids, p. 54

5.3.1.1 Haplocalcids

Key to Subgroups
Haplocalcids that have both:

1. An ECe of more than 8 to less than 30 dS m^{-1} in a layer 10 cm or more thick, within 100 cm of the soil surface: *and*
2. A lithic contact within 50 cm of the soil surface.

Salidic Lithic Haplocalcids

Other Haplocalcids that have a lithic contact within 50 cm of the soil surface (Fig. 5.3).

Lithic Haplocalcids

Fig. 5.3 Soilscape showing Lithic Haplocalcids (AD201)

Other Haplocalcids that have an ECe of more than 8 to less than 30 dS m^{-1} in a layer 10 cm or more thick, within 100 cm of the soil surface:

Salidic Haplocalcids

Other Haplocalcids that have, in a horizon at least 25 cm thick within 100 cm of the mineral soil surface, an exchangeable sodium percentage of 15 or more (or an SAR of 13 or more) during at least 1 month in normal years.

Sodic Haplocalcids

Other Haplocalcids that have a calcic horizon within 100 cm of the soil surface and lack all other characteristics of any other subsurface horizon (Fig. 5.4).

Typic Haplocalcids

Fig. 5.4 Soilscape showing Typic Haplocalcids (NE015)

5.3.1.2 Petrocalcids

Key to Subgroups

Petrocalcids that have a calcic horizon overlying the petrocalcic horizon.

Calcic Petrocalcids

Other Petrocalcids that have a petrocalcic horizon within 100 cm of the soil surface and lack all other characteristics of any other subsurface horizon (Fig. 5.5).

Typic Petrocalcids

Fig. 5.5 Soilscape showing Typic Petrocalcids (AD106)

5.3.2 Cambids

Key to Great Groups
The soils that are the most extensive of the Cambids and do not have aquic conditions (within 100 cm of soil surface), a duripan or a petrocalcic or petrogypsic horizon within 150 cm of the soil surface, and do not have an anthropic epipedon.

Haplocambids, p. 56

5.3.2.1 Haplocambids

Key to Subgroups
Haplocambids that have, in a horizon at least 25 cm thick within 100 cm of the soil surface, an exchangeable sodium percentage of 15 or more (or an SAR of 13 or more) during at least 1 month in normal years.

Sodic Haplocambids

Other haplocambids that have an irregular decrease in organic-carbon content (Holocene age) between a depth of 25 cm and either a depth of 125 cm below the mineral soil surface or a densic, lithic, or paralithic contact, whichever is shallower.

Fluventic Haplocambids

Other Haplocambids that lack all other characteristics (Fig. 5.6).

Typic Haplocambids

Fig. 5.6 Soilscape showing Typic Haplocambids (NE019)

5.3.3 Gypsids

Key to Great Groups

Gypsids that have a petrogypsic or petrocalcic horizon within 100 cm of the soil surface.

Petrogypsids, p. 61

Other Gypsids that have a calcic horizon within 100 cm of the soil surface.

Calcigypsids, p. 57

Other Gypsids that have only a gypsic horizon within 100 cm of the soil surface.

Haplogypsids, p. 59

5.3.3.1 Calcigypsids

Key to Subgroups

Calcigypsids that have a lithic contact within 50 cm of the soil surface (Fig. 5.7).

Lithic Calcigypsids

Other Calcigypsids that have an ECe of more than 8 to less than 30 dS m^{-1} in a layer 10 cm or more thick, within 100 cm of the soil surface (Fig. 5.8).

Salidic Calcigypsids

Fig. 5.7 Soilscape showing Lithic Calcigypsids (AD207)

Fig. 5.8 Soilscape showing Salidic Calcigypsids (AD232)

Other Calcigypsids that lack all other diagnostic characteristics. These soils represent the central concept of the great group.

Typic Calcigypsids

5.3.3.2 Haplogypsids

Key to Subgroups
Haplogypsids that have:

1. An ECe of more than 8 to less than 30 dS m^{-1} in a layer 10 cm or more thick, within 100 cm of the soil surface: *and*
2. A lithic contact within 50 cm of the soil surface.

Salidic Lithic Haplogypsids

Other Haplogypsids that have a lithic contact within 50 cm of the soil surface (Fig. 5.9).

Lithic Haplogypsids

Fig. 5.9 Soilscape showing Lithic Haplogypsids (AD215)

Other Haplogypsids that have:

1. An ECe of more than 8 to less than 30 dS m^{-1} in a layer 10 cm or more thick, within 100 cm of the soil surface: and
2. A gypsic horizon within 18 cm of the soil surface.

Salidic Leptic Haplogypsids

Other Haplogypsids that have a gypsic horizon within 18 cm of the soil surface (Fig. 5.10).

Leptic Haplogypsids

Fig. 5.10 Soilscape showing Leptic Haplogypsids (AD112)

Other Haplogypsids that have an ECe of more than 8 to less than 30 dS m^{-1} in a layer 10 cm or more thick, within 100 cm of the soil surface.

Salidic Haplogypsids

Other Haplogypsids that have only a gypsic horizon and lack all other characteristics. These soils represent the central concept of the great group (Fig. 5.11).

Typic Haplogypsids

Fig. 5.11 Soilscape showing Typic Haplogypsids (NE022)

5.3.3.3 Petrogypsids

Key to Subgroups

Petrogypsids that have both:

1. An ECe of more than 8 to less than 30 dS m^{-1} in a layer 10 cm or more thick, within 100 cm of the soil surface: *and*
2. A lithic contact within 50 cm of the soil surface.

Salidic Lithic Petrogypsids

Other Petrogypsids that have lithic contact within 50 cm of the soil surface.

Lithic Petrogypsids

Other Petrogypsids that have both:

1. An ECe of more than 8 to less than 30 dS m^{-1} in a layer 10 cm or more thick, within 100 cm of the soil surface: *and*
2. A calcic horizon overlying the petrogypsic horizon (Fig. 5.12).

Salidic Calcic Petrogypsids

Fig. 5.12 Soilscape showing Salidic Calcic Petrogypsids (AD119)

Other Petrogypsids that have a calcic horizon overlying the petrogypsic horizon (Fig. 5.13).

Calcic Petrogypsids

Fig. 5.13 Soilscape showing Calcic Petrogypsids (AD221)

Other Petrogypsids that have an ECe of more than 8 to less than 30 dS m^{-1} in a layer 10 cm or more thick, within 100 cm of the soil surface (Fig. 5.14).

Salidic Petrogypsids

Fig. 5.14 Soilscape showing Salidic Petrogypsids (AD227)

Other Petrogypsids that have only a petrogypsic horizon and lack all other diagnostic subsurface horizons or characteristics. These soils represent the central concept of the great group (Fig. 5.15).

Typic Petrogypsids

Fig. 5.15 Soilscape showing Typic Petrogypsids (AD226)

5.3.4 Salids

Key to Great Groups

Salids that are saturated with water in one or more layers within 100 cm of the mineral soil surface for 1 month or more in normal years.

Aquisalids, p. 63

Other Salids that are not saturated with water in one or more layers within 100 cm of the mineral soil surface for 1 month or more in normal years.

Haplosalids, p. 66

5.3.4.1 Aquisalids

Key to Subgroups

Aquisalids that have both:

1. A gypsic or petrogypsic horizon with its upper boundary within 100 cm of the soil surface, *and*

2. Lithic contact within 50 cm of the soil surface.

Gypsic Lithic Aquisalids

Other Aquisalids that have a lithic contact within 50 cm of the soil surface (Fig. 5.16).

Lithic Aquisalids

Fig. 5.16 Soilscape showing Lithic Aquisalids (NE 026)

Other Aquisalids that have a petrogypsic horizon with its upper boundary within 100 cm of the soil surface (Fig. 5.17).

Petrogypsic Aquisalids

Fig. 5.17 Soilscape showing Petrogypsic Aquisalids (AD143)

Other Aquisalids that have an anhydritic horizon with its upper boundary within 100 cm of the soil surface (Fig. 5.18).

Anhydritic Aquisalids

Fig. 5.18 Soilscape showing Anhydritic Aquisalids (AD125)

Other Aquisalids that have a gypsic horizon with its upper boundary within 100 cm of the soil surface (Fig. 5.19).

Gypsic Aquisalids

Fig. 5.19 Soilscape showing Gypsic Aquisalids (NE025)

Other Aquisalids that only have a salic horizon and do not have lithic contact within 50 cm of the soil surface or an anhydritic, calcic, gypsic, petrocalcic, or petrogypsic horizon that has its upper boundary within 100 cm of the soil surface. These soils represent the central concept of the great group (Fig. 5.20).

Typic Aquisalids

Fig. 5.20 Soilscape showing Typic Aquisalids (AD128)

5.3.4.2 Haplosalids

Key to Subgroups

Haplosalids that have both:

1. An overlying petrogypsic horizon: *and*
2. A lithic contact within 50 cm of the soil surface (Fig. 5.21).

Petrogypsic Lithic Haplosalids

Other Haplosalids that have *both*:

1. An overlying gypsic horizon: *and*
2. A lithic contact within 50 cm of the soil surface (Fig. 5.22).

Gypsic Lithic Haplosalids

Fig. 5.21 Soilscape showing Petrogypsic Lithic Haplosalids (AD235)

Fig. 5.22 Soilscape showing Gypsic Lithic Haplosalids (AD234)

Other Haplosalids that have *both*:

1. An overlying calcic horizon: *and*
2. A lithic contact within 50 cm of the soil surface (Fig. 5.23).

Calcic Lithic Haplosalids

Fig. 5.23 Soilscape showing Calcic Lithic Haplosalids (AD230)

Other Haplosalids that have a lithic contact within 50 cm of the soil surface (Fig. 5.24).

Lithic Haplosalids

Fig. 5.24 Soilscape showing Lithic Haplosalids (AD239)

Other Haplosalids that have a petrogypsic horizon within 100 cm of the soil surface (Fig. 5.25).

Petrogypsic Haplosalids

Fig. 5.25 Soilscape showing Petrogypsic Haplosalids (AD236)

Other Haplosalids that have an anhydritic horizon within 100 cm of the soil surface.

Anhydritic Haplosalids

Other Haplosalids that have a gypsic horizon within 18 cm of the soil surface.

Leptic Gypsic Haplosalids

Other Haplosalids that have a gypsic horizon at a depth of more than 18 cm to less than 100 cm from the soil surface (Fig. 5.26).

Gypsic Haplosalids

Fig. 5.26 Soilscape showing Gypsic Haplosalids (AD110)

Other Haplosalids that have a calcic horizon within 100 cm of the soil surface.

Calcic Haplosalids

Other Haplosalids that have only a salic horizon. These soils represent the central concept of the great group (Fig. 5.27).

Typic Haplosalids

Fig. 5.27 Soilscape showing Typic Haplosalids (AD145)

5.4 Key to the Suborders of Entisols

Entisols that have permanent saturation with water and a reduced matrix in all horizons below 25 cm from the mineral soil surface.

Aquents, p. 71

Entisols that have less than 35 % (by volume) rock fragments and a texture class of loamy fine sand or coarser in all layers (sandy loam lamellae are permitted) within the particle-size control section.

Psamments, p. 74

Other Entisols that occur on recent erosional surfaces. As such they typically represent soils that display a high content of rock fragments or, as in some cases in the UAE, shell fragments. The soils are sandy or loamy in texture, are well drained, and in some places overlie calcified bedrock.

Orthents, p. 71

5.4.1 Aquents

Key to Great Groups

Aquents that have less than 35% (by volume) rock fragments and a textural class of loamy fine sand or coarser in all layers (sandy loam lamellae are permitted) within the particle-size control section.

Psammaquents, p. 71

5.4.1.1 Psammaquents

Key to Subgroups

Psammaquents that have an ECe of more than 8 to less than 30 dS m^{-1} in a layer 10 cm or more thick, within 100 cm of the soil surface.

Salidic psammaquents

Psammaquents that represent central concept of great group, and lack the characteristics of other psammaquents (Lithic, sodic, spodic, humaqueptic, mollic).

Typic psammaquents

5.4.2 Orthents

Key to Great Groups

Orthents that have an aridic (or torric) soil moisture regime.

Torriorthents, p. 71

5.4.2.1 Torriorthents

Key to Subgroups

Torriorthents that have a lithic contact within 50 cm of the soil surface and an ECe of more than 8 to less than 30 dS m^{-1} in a layer 10 cm or more thick, within 100 cm of the soil surface (Fig. 5.28).

Salidic Lithic Torriorthents

Other Torriorthents that have a lithic contact within 50 cm of the soil surface (Fig. 5.29).

Lithic Torriorthents

Other Torriorthents that have an ECe of more than 8 to less than 30 dS m^{-1} in a layer 10 cm or more thick, within 100 cm of the soil surface (Fig. 5.30).

Salidic Torriorthents

Fig. 5.28 Soilscape showing Salidic Lithic Torriorthents (AD240)

Fig. 5.29 Soilscape showing Lithic Torriorthents (AD150)

Other Torriorthents that have, in a horizon at least 25 cm thick within 100 cm of the mineral soil surface, an exchangeable sodium percentage of 15 or more (or an SAR of 13 or more) during at least 1 month in normal years.

Sodic Torriorthents

Fig. 5.30 Soilscape showing Salidic Torriorthents (NE005)

Other Torriorthents that lack all other diagnostic characteristics. These are the soils that represent the central concept of the great group. They are typically moderately deep or deep, well drained, gravelly soils (Fig. 5.31).

Typic Torriorthents

Fig. 5.31 Soilscape showing Typic Torriorthents (AD151)

5.4.3 Psamments

Key to Great Groups
Psamments that have an aridic (or torric) soil moisture regime and do not have moisture available for plants for long periods.

Torripsamments, p. 74

5.4.3.1 Torripsamments

Key to Subgroups
Torripsamments that have *both*:

1. An ECe of more than 8 to less than 30 dS m^{-1} in a layer 10 cm or more thick, within 100 cm of the soil surface: *and*
2. A lithic contact within 50 cm of the soil surface.

Salidic Lithic Torripsamments

Other Torripsamments that have a lithic contact within 50 cm of the soil surface (Fig. 5.32).

Lithic Torripsamments

Fig. 5.32 Soilscape showing Lithic Torripsamments (AD152)

Other torripsamments that are saturated with water in one or more layers within 150 cm of the mineral soil surface in normal years for *either or both*:

1. 20 or more consecutive days; *or*
2. 30 or more cumulative days.

Oxyaquic Torripsamments

Other Torripsamments that have an ECe of more than 8 to less than 30 dS m^{-1} in a layer 10 cm or more thick, within 100 cm of the soil surface (Fig. 5.33).

Salidic Torripsamments

Fig. 5.33 Soilscape showing Salidic Torripsamments (AD246)

Other torripsamments that have, in a horizon at least 25 cm thick within 100 cm of the mineral soil surface, an exchangeable sodium percentage of 15 or more (or an SAR of 13 or more) during at least 1 month in normal years.

Sodic Torripsamments

Other Torripsamments that lack all other diagnostic characteristics. These are the soils that represent the central concept of the great group. They are deep and have a texture of sand or loamy sand throughout the top 100 cm of the profile. They are nonsaline and are always calcareous to various degrees but do not have enough pedogenic carbonate accumulation to form a calcic horizon. They are the most extensive soils in the United Arab Emirates (Fig. 5.34).

Typic Torripsamments

Fig. 5.34 Soilscape showing Typic Torripsamments (NE011)

Reference

Soil Survey Staff. (2010). *Keys to soil taxonomy* (11th ed.). Washington, DC: U.S. Department of Agriculture, Natural Resources Conservation Service, U.S. Government Printing Office.

Investigation of shallow water table and strongly crusted saline scald (aquisalids) in the coastal area

Laboratory soil analysis supports final establishment of soil taxa

Laboratory Methods for Classification of United Arab Emirates Soils

6

Abstract

Laboratory analyses of soil samples are necessary to verify field soil taxonomic observation. The standard laboratory methods (physical, chemical, engineering, mineralogical, and micromorphological) that form the basis of the operational definitions of the second edition of Soil Taxonomy are described in this chapter, and these methods have been used to support the United Arab Emirates Keys to Soil Taxonomy. Some modifications have been made to fit the hot desert conditions in the United Arab Emirates. The procedures of various determinations such as soil texture, fragments in the soil, moisture content, total pretreatment loss (TPL), loss on acid treatment, carbonate equivalents, gypsum, anhydrite, extractable cations, cation-exchange capacity, exchangeable sodium percentage, saturation percentage, saturation extract analysis, electrical conductivity of the saturated extract, soil reaction or hydrogen ion activity (pH), sodium adsorption ratio, osmotic potential, engineering data (Atterberg limits), AASHTO group classification, Unified Soil Classification System (USCS), percent passing sieves, water retention, bulk density, particle density, porosity, organic carbon, soil mineralogy, clay mineralogy, x-ray fluorescence, thin section study (soil micromorphology) are described in this chapter.

Keywords

Laboratory procedures • Physical • Chemical • Mineralogical • Engineering • Micromorphological

S.A. Shahid et al., *United Arab Emirates Keys to Soil Taxonomy*,
DOI 10.1007/978-94-007-7420-9_6,

6.1 Introduction

Laboratory analyses of soil samples are necessary to verify field observation data, to determine properties and characteristics that cannot be estimated accurately by field observations, to help characterize typical profiles, and to properly classify soils. Laboratory procedures are routinely checked and closely monitored to ensure quality control. The standard laboratory methods that form the basis of the operational definitions of the second edition of *Soil Taxonomy* (Soil Survey Staff 1999) are described in the *Soil Survey Laboratory Methods Manual* (Burt 2004) and have been used to support the *United Arab Emirates Keys to Soil Taxonomy*. Some modifications have been made to fit the hot desert conditions in the United Arab Emirates. The following analyses (physical, chemical, engineering, mineralogical, and micromorphological) were carried out. Method codes cited are taken from Burt (2004).

6.1.1 Soil Texture

The texture of a soil is defined according to percentages of sand, silt, and clay in the fraction of the soil that is less than 2 mm in diameter (fine-earth fraction). After pretreatments to remove organic matter and gypsum and to disperse the soil:water suspension, the soil texture was assessed by determining the particle-size distribution. Very coarse (1–2 mm), coarse (0.5–1 mm), medium (0.25–0.5 mm), fine (0.1–0.25 mm), and very fine (0.05–0.1 mm) sands were determined by wet sieving and oven drying the sand fractions retained on each sieve. Coarse silt (0.02–0.05 mm), fine silt (0.002–0.02 mm), and clay (<0.002 mm) were determined from an oven-dried aliquot of suspended soil collected by pipette at specified timings of sedimentation based on Stoke's Law. Results are presented as weight percentages of the fraction less than 2 mm on an oven-dry soil basis, including carbonates and excluding gypsum. Using the proportions of the sand, silt, and clay fractions, the soil texture class was determined using USDA specifications (Soil Survey Division Staff 1993).

6.1.2 Fragments in the Soil

The fraction which is 2 mm or larger in diameter are fragments in the soil. These fragments include rock fragments (i.e., geologic origin), shells, and carbonate nodules, but exclude gypsum crystals, and were determined by a dry sieving procedure. The weights of the material that would not pass sieves with apertures of 5, 20, and 75 mm were recorded.

6.1.3 Moisture Content

A known quantity of air-dried sample was weighed, dried to a constant weight in an oven at 110 ± 5°C, and reweighed. Moisture content of gypsiferous soils that are reported on an oven-dry weight basis was adjusted to remove the weight of the crystal water (methods 2D1-3 and 3D1-3).

6.1.4 Total Pretreatment Loss (TPL)

Carbonates were dissolved by heating the soils to approximately 90 °C in a NaOAc solution (buffered to pH 5). The undissolved soil was then separated from the NaOAc solution, and the soil sample treated with H_2O_2 for oxidation of organic matter. Distilled water was then added to remove soluble gypsum. The residual sample was oven-dried overnight. The loss in sample weight (due to removal of carbonates, organic matter, and gypsum) is calculated and reported as a percentage on a < 2 mm basis. This method is based on method 3A1a1a (steps 7.5–7.10) and method 3A1a2a

6.1.5 Loss on Acid Treatment (LAT)

A known quantity of sample was treated with HCl to destroy carbonates. The loss in weight is expressed as calcium carbonate on an oven-dry basis, <2 mm basis (based on method 6E1c). The data are a useful check on the accuracy of the carbonate equivalents analysis.

6.1.6 Calcium Carbonate Equivalents

The quantity of calcium carbonate equivalent is required in soil taxonomy to identify a calcic diagnostic horizon and to determine anhydritic, carbonatic, and gypsic mineralogy classes. The amount of carbonate in the soil was measured by treating the samples with HCl. The evolved CO_2 was measured manometrically using a calcimeter. The amount of carbonate was then calculated as percent $CaCO_3$ equivalent regardless of the form of carbonates (dolomite, sodium carbonate, magnesium carbonate, etc.) in the sample (method 4E1a1a1a1).

6.1.7 Gypsum

Gypsum ($CaSO_4.2H_2O$) content of a soil is a criterion for gypsic and petrogypsic horizons and for particle-size and mineralogical classes at the family level (Soil Survey Staff 1999). Gypsum was measured by reacting a soil extract with acetone, the precipitated calcium sulfate was dissolved in water and EC was measured enabling gypsum percentage to be derived from a calibration curve (method 4E2a1a1a1).

6.1.8 Anhydrite

Anhydrite ($CaSO_4$) is quantified by the difference in two analytical procedures. Anhydrite and gypsum are both extracted and measured by the acetone procedure (method 4E2a1a1a1). Gypsum (but not anhydrite) is quantified by thermal gravimetric analysis, a method that measures the weight loss of a sample by heating it from 20 °C to 200 °C at a rate of 2 °C per minute. The weight of water lost between 75 °C and 115 °C is used to quantify the gypsum based on a theoretical weight loss of 20.9 % (Karathanasis and Harris 1994). Finally, the percent anhydrite is equal to the difference between the acetone and thermal procedures.

6.1.9 Extractable Cations

The soil sample was reacted with 1 N NH_4OAc to replace exchangeable cations. The solution was then separated through centrifugation, and the extracted Na, K, Ca, and Mg were determined using Inductively Coupled Plasma-Atomic Emission Spectrometry (ICP-AES) and expressed as meq/100 g (Thomas 1982). For SI conversion the units can be considered as 1 meq/100 g = 1 cmol(+)/kg.

6.1.10 Cation-Exchange Capacity (CEC)

Cation-exchange capacity is a measure of the ability of a soil to hold and exchange cations in a plant available form (e.g., Ca, Mg, Na, NH_4, and K). It is one of the most important chemical properties of soil. This method was based on saturation of cation-exchange sites with Na by "equilibration" of the soil with 1 N NaOAc solution buffered at pH 8.2 and washing with 95 % ethanol. The extraction of sodium was carried out with 1 N ammonium acetate solution (pH 7). The extracted Na was then determined using ICP-AEC and expressed as the CEC in meq/100 g

(USDA 1954). For SI conversion the units can be considered as 1 meq/100 g = 1 cmol(+)/kg.

6.1.11 Exchangeable Sodium Percentage (ESP)

Exchangeable sodium percentage was calculated using the relationship given by the U.S. Salinity Laboratory Staff (USDA 1954) for estimating ESP of saturation extract from SAR for soils with CEC less than 50 meq/100 g: $[100 \times (-0.0126+0.01475 \times \mathrm{SAR})/(1+(-0.0126+0.01475 \times \mathrm{SAR})]$.

6.1.12 Saturation Percentage (SP)

The saturation percentage provides indication of available water contents in soil and is also a prerequisite for converting soil solution chemistry data (in water) to soil weight basis. A 250 g air-dried soil sample was saturated with water, taking care to ensure that the resulting soil/water mix met the criteria for a saturated paste. The volume of water and weight of soil were used to calculate the saturation percentage (SP) (method 4 F2).

6.1.13 Saturation Extract Analysis

The solution in the saturated paste was separated from the soil by vacuum extraction and filtration and analyzed for the following characteristics (methods 4F2c1a1 to 4): soluble Ca^{2+}, Mg^{2+}, Na^+, and K^+ were measured by Inductively Coupled Plasma-Atomic Emission Spectrometry (ICP-AES). Soluble HCO_3^- by titration with 0.011 N H_2SO_4 in the presence of methyl orange indicator, CO_3^{2-} by titration with 0.011 N H_2SO_4 in the presence of phenolphthalein indicator, and soluble Cl^-, NO_3^-, PO_4^{3-}, and SO_4^{2-} by ion chromatography.

6.1.14 Electrical Conductivity of the Saturated Extract (ECe)

The ECe is a standard representation of an indirect measurement of soil salinity (soluble salts commonly composed of contain Ca^{2+}, Mg^{2+}, Na^+, and K^+, HCO_3^-, CO_3^{2-}, Cl^-, and SO_4^{2-}). The ECe is measured with a standard EC meter and reported as deci Siemens/meter (dS m^{-1}).

6.1.15 Soil Reaction or Hydrogen Ion Activity (pH)

The soil reaction is an expression of the degree of acidity or alkalinity of a soil and is expressed as pH. The pH of saturated soil paste (pHs) is measured with a standard pH meter calibrated using buffer solutions of pH 4.0, 7.0, and 10.0.

6.1.16 Sodium Adsorption Ratio (SAR)

The SAR is a measure of the activities of sodium relative to calcium and magnesium in the soil saturation extract. It is an indirect measure of the equilibrium relation between sodium in the salt solution and exchangeable sodium adsorbed on the soil exchange complex. The SAR was calculated by standard formula (method 4F3b): $SAR = Na/[(Ca + Mg)/2]^{0.5}$, using inputs for the water soluble cations expressed as milliequivalents per liter (meq L^{-1}), and SAR expressed as (mmoles $L^{-1})^{0.5}$.

6.1.17 Osmotic Potential (OP)

Osmotic potential (atmospheres) provides a guide to the energy a plant must expend to extract water from soil solution; a more saline solution requires more energy. The ECe value was used to estimate the osmotic potential in atmospheres of a solution (USDA 1954) as $OP \approx 0.36 \times ECe$ where ECe is expressed as dS m^{-1}.

6.1.18 Engineering Data

The soils can be analyzed for their Atterberg limits and classified for engineering purposes according to the Unified Soil Classification System (USCS) and the system adopted by the American Association of State Highway and Transportation Officials (AASHTO).

6.1.19 Atterberg Limits

Atterberg limits are determined on the fraction less than 0.4 mm in size. Atterberg limits are soil water contents that define the upper and lower boundaries of the plastic state for a given soil. They provide basic measures of the engineering behavior of fine-grained soil and include the plastic limit (Wp) and liquid limit (Wl) and plasticity index (Ip), which is the difference in water content between the liquid limit and plastic limit. The plastic limit is determined by rolling a soil sample into 3.2 mm diameter thread. This thread is broken into pieces, squeezed together again, and rolled until its water content is reduced to a point at which the thread crumbles and can no longer to be pressed and rolled. At this point the water content of soil is reported as the plastic limit. Liquid limit is determined by performing observations in which a portion of sample is spread in a brass cup, divided in two by a grooving tool, and then allowed to flow together by repeatedly dropping the cup in a uniform manner in the standard mechanical liquid limit device. The multipoint liquid limit procedure requires the performance of three or more observations over a range of water contents and the plotting of data from the trials to obtain the relationship from which the liquid limit is determined (ASTM D 4318–2004).

6.1.20 AASHTO Group Classification

The AASHTO group classification is based upon particle-size distribution and Atterberg limits. The system is used for geotechnical engineering purposes to make general interpretations relating to the engineering performance of soils for construction and load bearing purposes.

The system identifies two general groups:

- Granular materials having 35 % or less, by weight, particles smaller than 0.074 mm in diameter; and
- Silty-clay materials having more than 35 %, by weight, particles smaller than 0.074 mm in diameter.

These two groups are further subdivided into seven main subgroup classifications. The group and subgroup classifications are based on estimated or measured grain-size distribution and on liquid limit and plasticity index values derived from measured values of samples taken from typical profiles.

The AASHTO group and subgroup classifications may be further modified by the addition of a group index value. The empirical group index formula was devised for approximate within-group evaluation of "clayey granular materials" and "silty-clay" materials. The dominant AASHTO group classifications for UAE soils and their inherent characteristics are:

A-3: Silty sand with a maximum of 10 % silt. Typically nonplastic. These soils are suitable for structural and general backfill material.

A-2-4: Silty sand with a maximum silt content of 35 %. The material is either nonplastic or has a low plasticity. The high sand content allows the soils to be classified as structural fill material and hence used for general engineering purposes (herein defined as construction and earthworks purposes).

A-1-b: These soils contain stone fragments and up to 25 % fines (very fine sand, silt, and clay). They are either nonplastic or slightly plastic. They are suitable as general backfill for engineering purposes.

A-1-a: These soils are similar to those classified as A-1-b but are limited by the high percentage of stone fragments.

6.1.21 Unified Soil Classification System (USCS)

The USCS is a system for classifying mineral and organic soils for engineering purposes based on particle-size characteristics, the amounts of various sizes, and the characteristics of the very fine particles assessed by the liquid limit and plasticity index.

The USCS identifies three major soil divisions:

(i) coarse-grained soils having less than 50 %, by weight, particles smaller than 0.074 mm in diameter;
(ii) fine-grained soils having 50 % or more, by weight, particles smaller than 0.074 mm in diameter; and
(iii) highly organic soils that demonstrate certain organic characteristics.

These divisions are further subdivided into 15 basic soil groups. The major soil divisions and basic soil groups are determined on the basis of estimated or measured values for grain-size distribution and Atterberg limits. The various groupings of the USCS classification have been devised to correlate with the engineering behavior of soils. This correlation provides a useful first step in any field or laboratory investigation of soils for engineering purposes. The classification and its interpretation are applicable in the design of foundations for a range of structures. The dominant USCS classifications for UAE soils and their inherent characteristics are:

SM: Silty sand with a maximum of 30 % fine material that is nonplastic. These soils contain various cementing agents, such as calcium carbonate, gypsum, clay, and silt. The degree of cementation can range from noncemented to strongly cemented in both the horizontal and vertical directions (note that within the soil profile various horizons of SM material can exhibit differing degrees of cementation). From an engineering point of view, the soil would classify as SM.

SP-SM: Silty sand with fine material limited to between 5 % and 12 %. These soils contain various cementing agents, such as calcium carbonate, gypsum, clay, and silt.

The soils are relatively permeable due to poor gradation. These soils are suitable for general engineering purposes.

SP: Poorly graded or gap graded fine to medium sand with little or no fines; not recommended for engineering purposes; difficult to compact; suitable for highly permeable fill material. This is the most common category for UAE soils.

SW: Well graded siliceous sand with little or no fine material; very good material for construction and general engineering purposes. Rarely found in the study area.

CL: Inorganic clay. Typically mixed with sands with low plasticity and Wl <50 %. These soils are not common in the UAE. This material is considered unsuitable for general engineering purposes.

MH: Inorganic silts with high plasticity. Very rare in the UAE. This material is considered unsuitable for general engineering purposes.

CH: Inorganic clays with high plasticity. Very rare in the UAE. This material is considered unsuitable for general engineering purposes.

6.1.22 Percent Passing Sieves

An air-dried soil subsample can be sieved through a nest of sieves, mesh numbers 4 (4.75 mm), 10 (2 mm), 40 (0.42 mm), and 200 (0.074 mm), and shaken mechanically. The material retained on each sieve can be weighed, and calculations can be made using a standard formula (USDA-NRCS 1995) to represent percent soil passing the respective sieves.

6.1.23 Water Retention (WR)

Water retention is defined as the soil water content at a given soil water suction. By varying the soil suction and recording the changes in soil water content, a water retention function or curve is determined. A pressure membrane apparatus method (3C1ae) is used to determine water retention (percent) at 1/10, 1/3, and 15 bars, respectively, for sieved, <2 mm, air-dry soil samples. The following are SI conversions:

15 bar = 1,500 kPa
1/3 bar = 33 kPa
1/10 bar = 10 kPa

6.1.24 Water Retention Difference (WRD)

The calculation of the WRD is considered the initial step in the approximation of the available water capacity. The WRD can be calculated as the difference in water retentions at 33 and1,500 kPa, converted to units of mm by multiplying by bulk density and to a whole soil basis by multiplying by the proportion of soil that passes a 2 mm sieve (method 3D5a).

6.1.25 Bulk Density (BD)

Bulk density is the weight of soil (oven dry) per unit bulk volume of soil (volume occupied by soil solids and pore spaces). The bulk density of a soil indicates the pore space available for water and roots. For example, a bulk density of 1.6 g/cm^3 and higher can restrict water storage and root penetration. Bulk density samples can be collected in standard steel cores of known volume and then oven dried. The loss in weight on drying is recorded and bulk density calculated by standard formula (method 4A3a). Where bulk density is not measured but is required for calculating other properties, an estimated value (BDcalc) is derived from the sand content (TOTALSAND%) based on the following equation derived from UAE soil surveys data: $\text{BDcalc} = 0.0098 \times \text{TOTALSAND\%} + 0.657$

6.1.26 Particle Density (PD)

The particle density provides an understanding of the basic component particles of the soil. The method requires using a pycnometer to estimate the volume of water or an inert gas (e.g., He) displaced by a known weight of soil, ensuring that all air is removed from the mixture (method 3G). Where particle density has not been measured, a value of 2.65 g/cm^3 has been assumed as the basis of other calculations.

6.1.27 Porosity

The percent volume of bulk soil that is occupied by pore spaces (the space filled by air or water) is called porosity. Porosity is calculated by using a standard formula of $100\,(1 - \text{BD}/\text{PD})$, where BD and PD represent the bulk density and particle density, respectively.

6.1.28 Organic Carbon (OC)

Organic carbon was determined by oxidizing the sample by flash combustion followed by reduction of gases produced. The product gases formed by reduction are separated chromatographically on a column using helium as carrier gas and detected using thermal conductivity detector (TCD). The percent organic carbon is converted to percent organic matter by multiplying by a factor of 1.724 (Burt 2004).

6.1.29 Soil Mineralogy (XRD Analysis)

The x-ray diffraction (XRD) is by far the most powerful technique to identify mineral species in soils. Whole soil mineralogy from the control sections of the typical soil profiles of identified soil families was determined by using a standard x-ray diffraction technique. The soil material was ground with an agate pestle and mortar and lightly pressed in an aluminum sample holder.

X-ray patterns were recorded using Cu-Kα radiation, using variable divergent, receiving and scattering slits. The step size was 1°2θ/min. The procedure is similar to method 7A2i (Burt 2004). A Philips x-ray diffraction model PW/1840, with Ni filter, Cu-Kα radiation ($\lambda = 0.154$ nm) at 40 kV, 55 mA, and scanning speed 0.02°/s, was used. Diffraction peaks between 2 and 60°2θ were recorded. The corresponding d-spacing and the relative intensities (I/Io) were calculated and compared with standard patterns (ASTM cards).

6.1.30 Clay Mineralogy (CM)

Clay (particles <2 μm) is a reactive (high surface area) soil component that controls physical, chemical, and nutritional behavior (nutrient storage) of soil and therefore affects agricultural potential. The clay-size material was separated from soil after the cementing material (organic matter, gypsum, and soluble salts) was removed for dispersion of clay in a soil suspension. Sedimentation and sampling for clay was based on Stoke's Law.

To resolve the clay mineral species, the clay was treated with K, K-heating at 300 °C and 550 °C, Mg, Mg ethylene glycol, and Mg-glycerol solvation. The treated clay samples were mounted on glass slides, and x-ray diffraction patterns were recorded using XRD equipment as above. The XRD patterns were interpreted for mineral species, comparing with standard reference patterns and measurements (ASTM cards). An additional test was made to confirm the presence of kaolinite using the Lim test (Lim et al. 1980). In this test the 7.2 Å peak of first order kaolinite shifts to 11.2 Å; however, second order chlorite peak remained at 7.00 Å. Based on the peak heights of the minerals, area under the peaks, and sharpness, the amounts of each mineral was estimated semiquantitatively (%).

6.1.31 X-ray Fluorescence (XRF)

A 1 g soil sample was ignited at 1,100 °C and dissolved in 9 g of lithium tetraborate and lithium borate (66:34) and 0.5 g of lanthanum oxide flux to form a disc for x-ray fluorescence analysis. The information is presented as percentages of elemental oxide.

6.1.32 Thin Section Study (Soil Micromorphology)

Undisturbed soil samples were collected in standard Kubiena boxes. Samples were dried and impregnated with a mixture of resin, acetone, and a hardener. On setting, the impregnated blocks were cut into small sections with a diamond saw and lubricant. The face of the section was polished with aloxite grit (6 μm) and then with diamond paste (6 and 3 μm) with Hyprez fluid as lubricant. It was then cleaned ultrasonically with petroleum spirit to remove residue from pores. The polished side was then stuck with adhesive on a clean slide and clamped to allow the adhesive layer to cure. The excess of the block was cut off with a diamond saw. The section was then ground initially to approximately 50 μm thickness on a lapping and polishing machine, then to 30 μm by hand grinding. The quartz grains were viewed in the microscope to observe the onset of gray birefringent color under crossed polars. Final polishing was completed with 6 and 3 μm diamond pastes followed by ultrasonic cleaning. Thin sections were studied on a petrographic microscope fitted with a digital camera, and the different components (groundmass, microstructure, mineral composition, pedofeatures, and pore space) were described using the procedures of Bullock et al. (1985) through point counting. Photographs of important features were captured using a digital camera fitted to the petrological microscope.

References

Bullock, P., Fedoroff, N., Jongerius, A., Stoops, G., Tursina, T., & Babel, U. (1985). *Handbook for soil thin section description*. Albrighton: Waine Research.

Burt, R. (2004). *Soil survey laboratory methods manual*. Soil Survey Investigations Report No. 42, Version 4.0. Nebraska: U.S. Department of Agriculture, Natural Resources Conservation Service, National Soil Survey Center.

Karathanasis, A. D., & Harris, W. G. (1994). Quantitative thermal analysis of soil minerals. In J. E. Amonette & L. W. Zelazny (Eds.), *Quantitative methods in soil mineralogy* (pp. 360–411). Wisconsin: Soil Science Society of America Miscellaneous Publication SSSA.

Lim, C. H., Jackson, M. L., Koons, R. D., & Helmke, P. A. (1980). Kaolins: Sources of differences in cation exchange capacities and cesium retention. *Clays and Clay Minerals, 28*, 223–229.

Soil Survey Division Staff. (1993). *Soil survey manual* (U.S. Department of Agriculture Handbook, No. 18). Washington, DC: U.S. Government Printing Office.

Soil Survey Staff. (1999). *Soil taxonomy: A basic system of soil classification for making and interpreting soil surveys* (U.S. Department of Agriculture Handbook 2nd ed., No. 436). Washington, DC: U.S. Government Printing Office.

Thomas, G. W. (1982). Exchangeable cations. In A. L. Page, et al. (Eds.), *Methods of soil analysis, part 2: Chemical and microbiological properties* (Agronomy monograph No. 9, 2nd ed., pp. 159–166). Madison: American Society of Agronomy, Inc./Soil Science Society of America, Inc.

United States Department of Agriculture, Natural Resources Conservation Service (USDA-NRCS). (1995). *Soil survey laboratory information manual* (Soil survey investigations report, No. 45, Version 1.0). Nebraska: National Soil Survey Center.

USDA. (1954). *Diagnosis and improvement of saline and alkali soils* (U.S. Department of Agriculture Handbook, No. 60). Washington, DC: U.S. Department of Agriculture.

Sandy deserts with some vegetation cover are attraction to grazers

Soil Families and Soil Series of the United Arab Emirates

7

Abstract

The US Soil Taxonomy hierarchy constitutes six levels (Order-suborder-great group-subgroup-family-soil series). In the UAE soil survey, soil family in the extensive survey and soil series in the intensive survey are recognized. The name of soil family and a soil series is used as a soil map unit component. There are 62 soil families and 19 soil families established with the correlation of extensive survey of Abu Dhabi Emirate and Dubai Emirate soil survey, respectively. There are a total of 74 soil series established with the correlation and publication of the soil surveys of the Abu Dhabi Emirate and the Northern Emirates. Salidic subgroup has been developed for the United Arab Emirates Keys to Soil Taxonomy and is recognized for the first time in this publication. In this chapter a list of the current identified soil families and soil series in the UAE and their official classification based on this Key are reported.

Keywords

Taxonomy • Hierarchy • Soil family • Soil series • Salidic • UAE

7.1 Introduction

The US soil taxonomy hierarchy constitutes of six levels (Order-suborder-great group-subgroup-family-soil series). The soil family in extensive survey of Abu Dhabi (EAD 2009a) and soil series are the lowest category of the UAE soil classification system (EAD 2009b, 2012). The soil family, name of soil series or their phases are the most common reference term used in soil map unit names, based on the scale and order of soil survey. The name of soil family and a soil series is also the most common reference term used as a soil map unit component.

S.A. Shahid et al., *United Arab Emirates Keys to Soil Taxonomy*,
DOI 10.1007/978-94-007-7420-9_7, © Springer Science+Business Media Dordrecht (Outside the USA) 2014

Soil series are the most homogeneous classes in the system of soil taxonomy and are the most common tool for communicating information about a particular soil type. As a general guide, a new soil series differs appreciably in either morphology or composition, or both, from already defined soil series. Differences in relevant characteristics must be larger than what may be normal errors of observation or estimates.

There are 62 soil families (Abu Dhabi Emirate), 19 soil families (Dubai Emirate) established with the correlation of extensive survey of Abu Dhabi Emirate (EAD 2009a) and Dubai Emirate soil survey (DM 2005) respectively (Tables 7.1 and 7.2). There are 74 soil series (Tables 7.3 and 7.4) established with the correlation and publication of the soil surveys of the Abu Dhabi Emirate (EAD 2009b) and the Northern Emirates (EAD 2012). At the time of publication of the soil surveys," shallow" families were assigned to those soils in the suborder of Salids with lithic contact within 50 cm of the soil surface. With the development of a soil classification system dedicated to the soils of the UAE, appropriate family criteria have been developed to identify these soils in a "Lithic" subgroup. These modifications have been incorporated into this publication and are proposed for revision to the USDA *Keys to Soil Taxonomy.*

Also, for those soils that have saline conditions within the upper 100 cm, a "saline" phase was assigned to the soil series name in the Abu Dhabi soil survey to recognize the high salt content. "Salidic" subgroups have been developed for the *United Arab Emirates Keys to Soil Taxonomy* and are recognized for the first time in this publication.

The Tables 7.1, 7.2, 7.3, and 7.4 provide a list of the current identified soil families and soil series in the UAE and their official classification based on this Key. Classification provided in the published soil survey reports is based on the version of the *Keys to Soil Taxonomy* that was current at the time of publication and may differ from the classification provided in Tables 7.1, 7.2, 7.3, and 7.4.

The soil series that are identified as "saline phase" (e.g., Saruk Qufa, saline phase) will be re-correlated to new soil series names. The column for "Soil Code" refers to the published soil families and soil series identifier. For example AD201 is the reference code assigned to the first soil series correlated in the *Soil Survey of Abu Dhabi Emirate-Intensive Survey* and NE001 is the first soil series correlated in the *Soil Survey of the Northern Emirates.*

It is assumed that when the surveyor keyed out the soil in the United Arab Emirates at the family or soil series levels, he/she will then correlate the information with Tables 7.1, 7.2, 7.3, and 7.4, which are providing soil codes for soil families and soil series as appeared in Soil Survey of Abu Dhabi Emirate (EAD 2009a, b), Dubai Emirate (DM 2005) and Soil Survey of Northern Emirates (EAD 2012). These survey reports present description of typical profiles following by soil analyses results, soil and thematic maps, mapping units and their component soils, as well as their interpretation for many uses.

Table 7.1 Soil families of the extensive soil survey of Abu Dhabi Emirate correlated to UAE Keys to Soil Taxonomy

Subgroup	Particle Size Class	Mineralogy Class	Activity	Temperature Regime Class	Other	Phase	Soil No
Salidic Haplocalcids	Sandy	Carbonatic		Hyperthermic		Salidic	AD101
Typic Haplocalcids	Sandy	Carbonatic		Hyperthermic		Lithic	AD102
Typic Haplocalcids	Sandy	Mixed		Hyperthermic			AD103
Salidic Haplocalcids	Sandy	Mixed		Hyperthermic		Lithic	AD104
Typic Torriorthents	Sandy-skeletal	Mixed		Hyperthermic			AD105
Typic Petrocalcids	Sandy	Carbonatic		Hyperthermic	Shallow		AD106
Typic Haplogypsids	Sandy	Mixed		Hyperthermic		Salidic	AD107
Salidic Calcigypsids	Sandy	Mixed		Hyperthermic		Lithic	AD108
Typic Calcigypsids	Sandy	Mixed		Hyperthermic		Petrogypsic	AD109
Gypsic Haplosalids	Sandy	Gypsic		Hyperthermic			AD110
Leptic Haplogypsids	Sandy	Gypsic		Hyperthermic		Lithic	AD111
Leptic Haplogypsids	Sandy	Mixed		Hyperthermic			AD112
Leptic Haplogypsids	Sandy	Mixed		Hyperthermic		Lithic	AD113
Leptic Haplogypsids	Sandy	Mixed		Hyperthermic		Petrogypsic	AD114
Salidic Leptic Haplogypsids	Sandy	Mixed		Hyperthermic			AD115
Gypsic Lithic Haplosalids	Sandy	Gypsic		Hyperthermic			AD116
Salidic Haplogypsids	Sandy	Gypsic		Hyperthermic			AD117
Salidic Haplogypsids	Sandy	Gypsic		Hyperthermic		Petrogypsic	AD118
Salidic CalcicPetrogypsids	Sandy	Mixed		Hyperthermic			AD119
Typic Petrogypsids	Sandy	Gypsic		Hyperthermic	Shallow		AD120
Typic Petrogypsids	Sandy	Gypsic		Hyperthermic			AD121
Salidic Petrogypsids	Sandy	Gypsic		Hyperthermic			AD122
Typic Petrogypsids	Sandy	Gypsic		Hyperthermic			AD123
Typic Petrogypsids	Sandy	Mixed		Hyperthermic	Shallow		AD124
Anhydritic Aquisalids	Fine	Anhydritic		Hyperthermic			AD125
Gypsic Aquisalids	Sandy	Gypsic		Hyperthermic			AD126

(continued)

Table 7.1 (continued)

Subgroup	Particle Size Class	Mineralogy Class	Activity	Temperature Regime Class	Other	Phase	Soil No
Gypsic Aquisalids	Sandy	Mixed		Hyperthermic			AD127
Typic Aquisalids	Sandy	Carbonatic		Hyperthermic			AD128
Typic Aquisalids	Sandy	Mixed		Hyperthermic			AD129
Typic Aquisalids	Sandy	Mixed		Hyperthermic		Lithic	AD130
Gypsic Haplosalids	Sandy	Mixed		Hyperthermic			AD131
Gypsic Haplosalids	Coarse-loamy	Mixed	Active	Hyperthermic		Lithic	AD132
Gypsic Lithic Haplosalids	Loamy	Mixed	Superactive	Hyperthermic			AD133
Gypsic Haplosalids	Sandy	Gypsic		Hyperthermic		Aquic	AD134
Gypsic Haplosalids	Sandy	Mixed		Hyperthermic			AD135
Gypsic Haplosalids	Sandy	Mixed		Hyperthermic		Aquic	AD136
Gypsic Haplosalids	Sandy	Carbonatic		Hyperthermic		Lithic	AD137
Petrogypsic Haplosalids	Coarse-loamy	Gypsic		Hyperthermic			AD138
Petrogypsic Haplosalids	Loamy	Gypsic		Hyperthermic	Shallow		AD139
Petrogypsic Haplosalids	Loamy	Mixed		Hyperthermic	Shallow		AD140
Petrogypsic Haplosalids	Sandy	Gypsic		Hyperthermic			AD141
Petrogypsic Haplosalids	Sandy	Gypsic		Hyperthermic	Shallow		AD142
Petrogypsic Aquisalids	Sandy	Mixed		Hyperthermic			AD143
Petrogypsic Haplosalids	Sandy	Mixed		Hyperthermic	Shallow		AD144
Typic Haplosalids	Coarse-loamy	Mixed	Subactive	Hyperthermic			AD145
Typic Haplosalids	Sandy	Carbonatic		Hyperthermic		Aquic	AD146
Typic Haplosalids	Sandy	Mixed		Hyperthermic		Aquic	AD147

Typic Haplosalids	Sandy	Mixed	Hyperthermic	Lithic	AD148
Lithic Haplosalids	Sandy	Mixed	Hyperthermic		AD149
Lithic Torriorthents	Sandy-skeletal	Mixed	Hyperthermic		AD150
Typic Torriorthents	Sandy-skeletal	Mixed	Hyperthermic		AD151
Lithic Torripsamments		Carbonatic	Hyperthermic		AD152
Lithic Haplocalcids	Sandy	Mixed	Hyperthermic		AD153
Salidic Torripsamments		Carbonatic	Hyperthermic	Petrogypsic	AD154
Typic Torripsamments		Carbonatic	Hyperthermic	Calcic	AD155
Sodic Haplocalcids	Sandy	Carbonatic	Hyperthermic	Lithic	AD156
Salidic Torripsamments		Gypsic	Hyperthermic	Shelly	AD157
Typic Torripsamments		Mixed	Hyperthermic		AD158
Salidic Torripsamments		Mixed	Hyperthermic	Calcic	AD159
Salidic Torripsamments		Mixed	Hyperthermic	Lithic	AD160
Typic Torripsamments		Mixed	Hyperthermic	Petrocalcic	AD161
Salidic Haplocalcids	Sandy	Mixed	Hyperthermic	Petrogypsic	AD162

Table 7.2 Soil families of the soil survey of Dubai Emirate correlated to UAE Keys to Soil Taxonomy

Subgroup	Particle Size Class	Mineralogy Class	Temperature Regime Class	Other	Soil No[a]
Sodic Torripsamments		Carbonatic	Hyperthermic		DE1
Oxyaquic Torripsamments		Carbonatic	Hyperthermic		DE2
Oxyaquic Torripsamments		Carbonatic	Hyperthermic	Salidic	DE3
Salidic Torripsamments		Carbonatic	Hyperthermic		DE4
Typic Toripsamments		Carbonatic	Hyperthermic		DE5
Salidic Psammaquents		Carbonatic	Hyperthermic		DE6
Sodic Torriorthents	Sandy-skeletal	Mixed	Hyperthermic		DE7
Sodic Torriorthents	Sandy-skeletal	Carbonatic	Hyperthermic		DE8
Sodic Torriorthents	Loamy-skeletal	Mixed	Hyperthermic		DE9
Salidic Torriorthents	Sandy-skeletal	Mixed	Hyperthermic		DE10
Salidic Torriorthents	Sandy-skeletal	Carbonatic	Hyperthermic		DE11
Salidic Torriorthents	Loamy-skeletal	Mixed	Hyperthermic		DE12
Lithic Torriorthents	Loamy-skeletal	Mixed	Hyperthermic		DE13
Lithic Torriorthents	Sandy-skeletal	Mixed	Hyperthermic		DE14
Typic Torriorthents	Loamy-skeletal	Mixed	Hyperthermic		DE15
Typic Torriorthents	Sandy-skeletal	Mixed	Hyperthermic		DE16
Typic Aquisalids	Fine-loamy	Carbonatic	Hyperthermic		DE17
Fluventic Haplocambids	Loamy-skeletal	Mixed	Hyperthermic		DE18
Typic Haplocambids	Loamy-skeletal	Mixed	Hyperthermic		DE19

[a]Soil families are derived from Soil Series of Dubai Emirate Soil Survey

Table 7.3 Soil series and soil classification of the intensive soil survey of Abu Dhabi Emirate correlated to UAE Keys to Soil Taxonomy

Soil Series	Subgroup	Particle Size Class	Mineralogy Class	Cation Exchange Activity Class	Temperature Regime Class	Other	Phase	Soil No
Sha'bbiyat	Lithic Haplocalcids	Sandy	Mixed		Hyperthermic			AD201
Markiya	Lithic Calcigypsids	Loamy	Carbonatic		Hyperthermic			AD202
Riqayyib	Salidic Lithic Haplocalcids	Loamy	Mixed	Superactive	Hyperthermic			AD203
Ghadar	Typic Haplocalcids	Sandy	Carbonatic		Hyperthermic		Lithic	AD204
Al Mayann	Typic Haplocalcids		Mixed		Hyperthermic			AD205
Al Babah	Typic Haplocalcids	Coarse-loamy	Mixed	Superactive	Hyperthermic			AD206
Tarbush	Lithic Calcigypsids	Sandy	Mixed		Hyperthermic			AD207
Saruk Kufa	Typic Calcigypsids	Sandy	Mixed		Hyperthermic		Salidic	AD208
Saruk Kufa, saline phase	Salidic Calcigypsids	Sandy	Mixed		Hyperthermic		ECe between 3 & 10	AD209
Al Niddah	Typic Calcigypsids	Sandy	Mixed		Hyperthermic		Lithic	AD210
Ghuweifat	Leptic Haplogypsids	Sandy	Mixed		Hyperthermic		Petrogypsic	AD211
Ghuweifat, saline phase	Salidic Leptic Haplogypsids	Sandy	Mixed		Hyperthermic		Lithic between 175 & 200 cm	AD212
Jawaloodh	Leptic Haplogypsids	Sandy	Mixed		Hyperthermic		Lithic	AD213
Jawaloodh, saline phase	Salidic Leptic Haplogypsids	Sandy	Mixed		Hyperthermic		Lithic between 75 & 95 cm	AD214
Al Daghar	Lithic Haplogypsid	Sandy	Gypsic		Hyperthermic			AD215
Al Daghar, saline phase	Salidic Lithic Haplogypsids	Sandy	Gypsic		Hyperthermic			AD216
Sila	Gypsic Lithic Haplosalids	Loamy	Mixed	Superactive	Hyperthermic			AD217
Ghayathi	Typic Haplogypsids	Sandy	Mixed		Hyperthermic			AD218

(continued)

Table 7.3 (continued)

Soil Series	Subgroup	Particle Size Class	Mineralogy Class	Cation Exchange Activity Class	Temperature Regime Class	Other	Phase	Soil No
Ghayathi, saline phase	Salidic Haplogypsids	Sandy	Gypsic		Hyperthermic			AD219
Maydur	Salidic Haplogypsids	Sandy	Mixed		Hyperthermic		Lithic between 50 & 100 cm	AD220
Al Khabar	Calcic Petrogypsids	Sandy	Mixed		Hyperthermic		Salidic	AD221
Salabik	Typic Petrogypsids	Sandy	Gypsic		Hyperthermic	Shallow		AD222
Salabik, saline phase	Salidic Petrogypsids	Sandy	Gypsic		Hyperthermic	Shallow		AD223
Barakah	Typic Petrogypsids	Sandy	Gypsic		Hyperthermic	Shallow		AD224
Barakah, saline phase	Salidic Petrogypsids	Sandy	Mixed		Hyperthermic	Shallow		AD225
Ghutail	Typic Petrogypsids	Sandy	Mixed		Hyperthermic			AD226
Ghutail, saline phase	Salidic Petrogypsids	Sandy	Mixed		Hyperthermic			AD227
At Taf	Gypsic Aquisalids	Sandy	Mixed		Hyperthermic			AD228
Shuweihat	Typic Aquisalids	Sandy	Mixed		Hyperthermic			AD229
Ruwais	Calcic Lithic Haplosalids	Loamy	Mixed	Subactive	Hyperthermic			AD230
Al Ashtan	Gypsic Lithic Haplosalids	Sandy	Mixed		Hyperthermic			AD231
Bida Hazza'a	Salidic Calcigypsids	Sandy	Mixed		Hyperthermic		ECe 10 or more	AD232
Al Uquayiah	Gypsic Haplosalids	Sandy	Gypsic		Hyperthermic		Lithic	AD233

Baynunah	Gypsic Lithic Haplosalids	Loamy	Mixed	Superactive	Hyperthermic		AD234
Al Hamra	Petrogypsic Lithic Haplosalids	Sandy	Mixed		Hyperthermic		AD235
Tarif	Petrogypsic Haplosalids	Sandy	Mixed		Hyperthermic		AD236
Harmiya	Typic Haplosalids	Sandy	Mixed		Hyperthermic		AD237
Khaberah	Typic Haplosalids	Sandy	Mixed		Hyperthermic	Lithic between 65 & 95 cm	AD238
Al Abyad	Lithic Haplosalids	Loamy	Mixed	Active	Hyperthermic		AD239
Al Usb	Salidic Lithic Torriorthents	Sandy-skeletal	Mixed		Hyperthermic		AD240
Musharib	Salidic Torriorthents	Sandy-skeletal	Carbonatic		Hyperthermic		AD241
Al Ain	Typic Torriorthents	Sandy-skeletal	Mixed		Hyperthermic		AD242
Al Hiwah	Lithic Torripsamments		Mixed		Hyperthermic		AD243
Al Hiwah, saline phase	Salidic Lithic Torripsamments		Mixed		Hyperthermic		AD244
Madinat Zayed	Typic Torripsamments		Mixed		Hyperthermic		AD245
Madinat Zayed, saline phase	Salidic Torripsamments		Mixed		Hyperthermic		AD246
Al Zafrah	Typic Haplosalids	Sandy	Mixed		Hyperthermic	Lithic between 100 &130 cm	AD247

Table 7.4 Soil series and soil classification of soil survey of the Northern Emirates correlated to UAE Keys to Soil Taxonomy

Soil Series	Subgroup	Particle Size Class	Mineralogy Class	Cation Exchange Activity Class	Temperature Regime Class	Other	Phase (Differentiating Criteria)	Soil No
Shibekah	Salidic Torriortents	Sandy	Carbonatic		Hyperthermic	Aquic-shelly		NE001
Fujairah	Typic Torriorthents	Fragmental	Mixed		Hyperthermic			NE002
Al Ain	Typic Torriorthents	Sandy-skeletal	Mixed		Hyperthermic		ESP <15; cobles/ stones < 15%	NE003
Bih	Typic Torriorthents	Sandy-skeletal	Mixed		Hyperthermic		Cobles/stones ≥15%	NE004
Miliha	Salidic Torriortents	Sandy-skeletal	Mixed		Hyperthermic		ESP ≥15; cobles/ stones < 15%	NE005
As Sirer	Typic Torriorthents	Sandy	Mixed		Hyperthermic		< 35% gravels within 100cm	NE006
Kuderah	Typic Torriorthents	Sandy	Mixed		Hyperthermic		≥ 35% gravels within 100cm	NE007
Ar Riwedah	Typic Torriorthents	Coarse-loamy	Carbonatic		Hyperthermic		>1/2 PSCS[1] is finer than VFS	NE008
Ajman	Typic Torriorthents	Coarse-loamy	Carbonatic		Hyperthermic		PSCS[1] is dominated by VFS	NE009
Al Aswad	Typic Torriorthents	Coarse-loamy	Carbonatic		Hyperthermic		Human trans-ported material	NE010
Sharjah	Typic Torripsamments		Carbonatic		Hyperthermic		25-49% of total sand is VFS	NE011
Al Murrah	Typic Torripsamments		Carbonatic		Hyperthermic		< 25% of total sand is VFS	NE012
Digdaga	Sodic Haplocalcids	Coarse-loamy	Carbonatic		Hyperthermic			NE013
Al Dhaid	Typic Haplocalcids	Sandy-skeletal	Mixed		Hyperthermic		Total silt + clay <15%	NE014

Al Kabkub	Typic Haplocalcids	Sandy-skeletal	Mixed		Hyperthermic		Total silt + clay ≥ 15%	NE015
Al Maalla	Typic Haplocalcids	Sandy	Mixed		Hyperthermic			NE016
Al Kihef	Typic Haplocalcids	Sandy	Carbonatic		Hyperthermic			NE017
Khatt	Sodic Haplocambids	Coarse-loamy	Mixed	Active	Hyperthermic			NE018
Ras Al Khaimah	Typic Haplocambids	Coarse-silty	Carbonatic		Hyperthermic			NE019
As Sihebi	Typic Haplocambids	Sandy over loamy	Carbonatic		Hyperthermic			NE020
Wahala	Leptic Haplogypsids	Coarse-loamy over sandy or sandy skeletal	Gypsic over carbonatic		Hyperthermic			NE021
Al Yafar	Typic Haplogypsids	Loamy-skeletal	Mixed	Superactive	Hyperthermic			NE022
Al Madam	Typic Calcigypsids	Coarse-loamy	Carbonatic		Hyperthermic			NE023
Ar Ramlah	Typic Aquisalids	Sandy	Carbonatic		Hyperthermic			NE024
Um Al Quwain	Gypsic Aquisalids	Fine-gypseous over loamy	Hypergypsic over carbonatic		Hyperthermic			NE025
Ar Rafah	Lithic Aquisalids	Sandy	Carbonatic		Hyperthermic			NE026
Hisan	Gypsic Haplosalids	Sandy	Carbonatic		Hyperthermic	Aquic		NE027
Misekhin	Gypsic Haplosalids	Coarse-silty	Gypsic		Hyperthermic			NE028

Differentiating criteria are provided only where there is more than one series in a given family
[1]*FSCS* particle size control section, *VFS* very fine sand, *ESP* exchangeable sodium percentage

References

Dubai Municipality (DM). (2005). *Satellite imagery and thematic mapping project*. Executed jointly by Global Scan Technologies (Dubai) and National Remote Sensing Agency, p. 221.

EAD. (2009a). *Soil survey of Abu Dhabi Emirate-extensive survey* (Vol. 1, pp. xx+506). Environment Agency – Abu Dhabi, UAE.

EAD. (2009b). *Soil survey of Abu Dhabi Emirate-intensive survey* (Vol. 2, pp. xvii+435). Environment Agency – Abu Dhabi, UAE.

EAD. (2012). *Soil survey of the Northern Emirates* (Soil report, Vol. 1, pp. xxvi+410). Environment Agency – Abu Dhabi, UAE.

Index

S.A. Shahid et al., *United Arab Emirates Keys to Soil Taxonomy*,
DOI 10.1007/978-94-007-7420-9, © Springer Science+Business Media Dordrecht (Outside the USA) 2014

FSC
www.fsc.org

MIX
Papier aus verantwortungsvollen Quellen
Paper from responsible sources
FSC® C105338

If you have any concerns about our products,
you can contact us on
ProductSafety@springernature.com

In case Publisher is established outside the EU,
the EU authorized representative is:
Springer Nature Customer Service Center GmbH
Europaplatz 3, 69115 Heidelberg, Germany

Printed by Libri Plureos GmbH
in Hamburg, Germany

An Employer's Guide to the Employment Protection Act 1975

An Employer's Guide to the

Employment Protection Act 1975

E. J. and M. D. Robertson

KOGAN PAGE

First published 1976 by
Kogan Page Limited
116a Pentonville Road, London N1 9JN

Printed in Great Britain by The Anchor Press Ltd
and bound by Wm Brendon & Son Ltd
both of Tiptree, Essex

ISBN 0 85038 201 7

Contents

Part III — The Institutions

Preface

We have both, from the standpoint of our different jobs, followed the long slow process which has finally put the Employment Protection Act on to the Statute Book. We formed the conviction that a text like this, embracing the experience of an employers' organisation official who has seen his staff struggle conscientiously with the Department of Employment and the Parliamentary process over more than a year, with the practical experience of someone working in a busy staff personnel area, would be of practical assistance to employers.

We have not attempted to write a definitive legal text on the new Act. What we have tried to do is to interpret what we perceive the provisions of the Act to mean and to assess the implications for employers. Mr. Michael Foot, Secretary of State for Employment, said in introducing this Act that it would provide a basis of labour practice and law that would serve for generations to come. Leaving aside some of the controversial extension of rights of trade unions, employers may care to regard Mr. Foot's statement as accurate and assume that the Employment Protection Act will not only have to be studied by them in a defensive short-term manner, but in a constructive long-term way as well. It strikes us both as a piece of legislation which will be around for a very long time and, however much it may go against the grain, British employers will have to settle down to the obligations they incur because of it.

This book was written in the limited time available to two busy people. That it does not contain more errors than it might have done (and we, of course, take full responsibility for any that occur) is due to much patient work by Olga Aikin, Barrister-at-law, of the London Business School and Peter Paterson, journalist and author.

Special and particular thanks are due to Morag for looking after Amy on successive Sundays.

The authors would also like to extend their thanks to Miss V. Bolwell for her invaluable help in many practical ways and to the Confederation of British Industry, the Advisory, Conciliation and

Arbitration Service, the Central Offices of the Industrial Tribunals and to the Department of Employment for permission to reproduce material contained in the book as appendices.

It should be noted that the interpretations, views and opinions contained in this book are those of the authors and cannot necessarily be attributed to the CBI or the P & O Steamship Navigation Company for whom the authors, respectively, work. A word of thanks is due to W. O. Campbell Adamson, Director-General of the CBI, for his permission to let his Director of Industrial Relations publish this book.

Introduction: Coping with the Act

The Employment Protection Act is the most important and far-reaching piece of legislation ever to be enacted in this country in the fields of employee relations and collective bargaining. Every employer, whether or not he recognises trade unions, will be affected by it and will need to be aware of at least its broad implications. Only the larger companies will be able to nominate a specialist to study the Act and arrive at detailed plans of action to cope with it. For the rest, the most they will be able to do is to recognise the obligations placed upon them as employers, assess how their employees and unions will handle their new rights and make commonsense plans accordingly.

This book is aimed primarily at the average employer and manager who, whilst perhaps aware of the leisurely progress of the Employment Protection Act through Parliament, has on the whole been much too busy operating a business in difficult times to keep up to date with yet more legislation. But legislation in the employment field is increasing and its rate of increase is fundamentally changing old relationships between employers and employed. The sequence of the Contracts of Employment Act 1972, the Redundancy Payments Act 1965, the Race Relations Act 1968, the Equal Pay Act 1970, the Industrial Relations Act 1971, the Trade Union and Labour Relations Act 1974, the Health and Safety Act 1974, the Employment Protection Act 1975, the Sex Discrimination Act 1975, and the awaited Trade Union and Labour Relations (Amendment) Act, as well as intended legislation on industrial democracy, places many obligations upon employers.

ORIGINS OF THE EMPLOYMENT PROTECTION ACT

In July 1973, the Trades Union Congress/Labour Party Liaison Committee discussed what legal action would be required to repeal and replace the Industrial Relations Act. Even at that time, with

Labour in opposition and with the TUC in vigorous opposition to the Industrial Relations Act, the principle of replacing and not simply repealing that Act had been conceded. This was so because the 1971 Act contained provisions, such as a statutory right not to be unfairly dimissed, which were attractive to the unions.

Thus it was agreed in the summer of 1973 that the next Labour Government would introduce legislation to repeal the Industrial Relations Act but preserve those factors thought desirable by the unions. Thereafter the future government would introduce an 'Employment Protection Bill' and an 'Industrial Democracy Bill'. Upon the return of the minority Labour Government in February 1974 no time was lost, understandably in the precarious political situation, in introducing the Trade Union and Labour Relations Bill on 30th April 1974. So as to ensure a quick passage of this 'repeal' bill through Parliament it seemed pretty clear that the TUC had tacitly agreed to defer some things they ideally would have wished to see in the first Bill, to the second, i.e. the Employment Protection Bill. These are understood to have included changes in the law on picketing and the right to be reinstated in cases of unfair dismissal, as well as the setting up of the Advisory, Conciliation and Arbitration Service (ACAS) on a statutory basis. In the event, however, the proposed amendments on picketing were dropped from the Employment Protection Act.

The Employment Protection Act has been hotly attacked by employers' organisations, which is hardly surprising in view of its one-sided origins. The CBI, for example, in June 1975* in response to the question 'Are employers being reactionary in their opposition to the proposals?' stated 'No. These are entirely one-sided proposals, drawn up after extensive consultation with the TUC and minimal contact with employers' representatives. Trade unions call continuously for free collective bargaining. Yet they here support legislation in areas which have been traditionally free. Employers have a duty to oppose any proposals which will increase costs in the present grave circumstances. In any case, none of the proposals will help employers to resolve their problems at this time. Whilst it is true that some of the proposals are already applied by leading firms, it is not true that all employers can afford to emulate them.'

This statement has been quoted at length because it illustrates the dilemma posed by this legislation for employers' organisations. On

*'Talking Point' No. 7: *The Employment Protection Bill.*

the one hand it is perfectly evident that the provisions of the Act are loaded in favour of trade unions and employees; on the other hand it is really not possible to object to individual rights such as the right to guarantee pay when laid off or to maternity pay and leave *in principle*. It is, of course, perfectly sensible to object to the institution of these rights at this time and at the employers' expense, but this is another matter. The fact is that many large employers will scan the new Act and conclude that whilst it represents some oncost and will involve them in some tedious additional administration, it can, on the whole, be lived with. True, the statutory enactment of rights cuts down on the range of 'package' responses to trade union claims; there is no possibility of offering a mix of improvements in wages and concessions on individual rights and conditions if the conditions part of the package has been pre-empted by legislation.

The position of the small employer is however very different indeed; for him the new rights will constitute a serious burden – so serious that some of them may consider that the Employment Protection Act is the last straw and sell up. It may well be that in this important sector of employment that job opportunities will be reduced by the new Act and that the temptation to play games with contracts of employment and self-employment will be very much increased.

In broad terms the Employment Act will affect employers in two ways. First it will give employees a set of new rights; these new rights will of course be new obligations for employers who will have to study them carefully and work out how to handle them. Secondly the Act will give independent trade unions a whole set of new rights – to the union as such and to individual trade union members.

Thus this book devotes Part I (Chapters 1–3) to setting out the new individual rights so that employers may clearly understand what they are, and most importantly understand how to deal with them in simple terms.

Chapter 1 discusses the so-called 'time off' provisions. These cover the statutory right of workers to guarantee pay when laid off; payment for employees suspended for health or safety reasons; the important introduction of maternity pay and the right to return to work for pregnant employees; time off, and in certain cases payment, for employees pursuing trade union duties or activities, or public duties such as serving as a local councillor or as a member of an area health board, and, finally, the right to paid time to look for work when redundancy is proposed. A detailed summary of these new time off provisions completes this chapter. The summary highlights the various service and other qualifications which employees must have to entitle them to access to these rights, and gives points designed to assist employers to plan for the introduction of these novel provisions.

Chapter 2 examines the new rights extended to workers in connection with dismissal. Employers have grown accustomed to the operation of the law in relation to dismissals and claims for unfair dismissal. The new provisions make it imperative that employers act impeccably about dismissal. An employee whose claim that he has been unfairly dismissed is upheld by an industrial tribunal will now be asked if he wishes to be reinstated on the job from which he was dismissed, with all losses made up to him, or re-engaged on some other job which is comparable to the old one. Should the tribunal decide that either reinstatement or re-engagement is appropriate they will issue an order to that effect. The penalties on employers for non-compliance with such an order may be, in extreme cases, up to a full year's pay.

Chapter 2 also examines the obligation upon employers to furnish written reasons for dismissal if requested, and the new rights of employees not to be harassed by actions which stop short of dismissal by employers who object to their engaging in trade union membership or activities. (The basis of compensation for unfair dismissal is in fact considerably changed by the Employment Protection Act and the potentially expensive consequences of these changes for employers is assessed.) Crucially the right not to be unfairly dismissed is extended to those working for very small employers; previously employers who employed four workers or less or who employed only close relations – a situation common in agriculture or small retail shops – did not have to concern themselves with the provisions of the old Industrial Relations Act which dealt with unfair dismissal. Now they do, with modest exceptions, and the material contained in Chapter 2 is designed to get this important fact over to every small employer who may, at present, be totally unaware of these new rights possessed by their few employees. The chapter concludes with a description of some technical, but important, changes in the Redundancy Payments Act brought about by the passing of the Employment Protection Act. These changes cover not only the working of a trial period after a new job offer has been accepted by an employee who would otherwise be redundant, but also the extension of the time period in which an employee may complain to a tribunal to obtain redundancy compensation. The chapter concludes with a simple disciplinary procedure which has been devised to take account of the new dismissal provisions.

Chapter 3 concludes the examination of individual rights and their implications for employers by describing a mixture of innovations, such as new rules for itemised pay statements, and changes in existing laws such as the Contracts of Employment Act and the Health and Safety at Work Etc. Act.

The book then moves on, in Part II (Chapters 4–8) to changes which will extend trade union rights, and other changes which will have an effect upon collective bargaining. Chapter 4 describes the legal assistance to be given to unions trying to get recognition from employers; these provisions start with the role of the Advisory, Conciliation and Arbitration Service, and then deal with the further role of the Central Arbitration Committee and the sanctions which that new body can bring to bear against employers who resist an ACAS recommendation that they recognise a particular union.

Chapter 5 describes the somewhat analogous legal assistance to be given to unions to force employers to give them information about the business which the unions claim is necessary to enable them to do their job of collectively bargaining for employees.

Chapter 6 looks at the role of the ACAS in disputes and strikes to assess what the likely effect will be of this independent body upon these critical areas; the powers of ACAS in respect of conciliation, arbitration, advice-giving and inquiry undertaking are described. This chapter concludes with a look at some highly technical changes in entitlement to Social Security and supplementary benefits and draws the attention of employees to the implications of these changes which, in effect, make it easier for an employee affected by a strike, but not actually on strike himself, to claim benefit.

Chapter 7 examines the requirement of employers to consult with recognised independent unions about proposed redundancies and the parallel requirement to notify the Department of Employment of proposed redundancies affecting 10 or more employees over 60 days and 100 or more over 90 days – thus falling into line with proposals emanating from the European Economic Community.

Chapter 8 focuses attention on a complex change involving the repeal of Section 8 of the Terms and Conditions of Employment

Act, 1959, which will enable unions to use the Central Arbitration Committee to arbitrate on claims for local or industry pay norms. It may be noted that this provision has been strenuously opposed by employers, and is highly unlikely to be brought into effect while incomes policies – statutory or 'voluntary' – are in operation, though no doubt there will be union pressure to advance it. This chapter concludes with a description of changes affecting Wages Councils which will be of intense interest to the employers of the three million or so employees in these fields.

Part III of the book (Chapter 9) deals with the new courts set up under the Employment Protection Act; namely, the Employment Appeal Tribunal, which looks uncommonly like Sir John Donaldson's National Industrial Relations Court but in fact differs significantly from that short-lived body; at the role and duties of what the Act calls 'the Service', namely the Advisory, Conciliation and Arbitration Service; the Central Arbitration Committee, and that body's offshoot, the Certification Officer. This chapter on institutions also deals with the extensive powers given to the Secretary of State for Employment under the Act, powers, for example, to ask Parliament to revise the limits on compensation for unfair dismissals and redundancy. We also examine the greatly extended part that will be played in employment matters by the already established industrial tribunals and trace out the relationship between the institutions that specifically operate the new Act and existing legal institutions such as the Appeal Court.

The appendices to the book give employers additional information on ACAS and industrial tribunals. We also include a CBI guide to employers on the subject of information and details of the Temporary Employment Subsidy, which was brought in before the Act received Royal consent but which is an integral part of the new provisions.

The authors share the conviction that not only specialists in the employment field, but all employers have a long and hard job ahead of them to understand and apply the provisions of this new Act. We hope that this guide will enable them to start this job. The action plan which follows this introduction may help them make such a start.

Action Plan

1. The provisions of the Employment Protection Act will be introduced by stages. The more onerous financial obligations may not arrive for fifteen months or more after the Act became law in November 1975. Watch Government announcements for implementation, but the likely sequence will be:

1st phase	Advisory, Conciliation and Arbitration Service – statutory status Central Arbitration Committee – to be set up. Certification Officer – to be set up. Employment Appeal Tribunal – to be set up. Recognition procedures. Safety representatives (trade union). Rights in respect of trade union membership and activities. Wages Councils' provisions. Employment agencies; transfer of licensing.
2nd phase	Disclosure of information and Code of practice. Time off for union and public duties and Code of practice. Contracts of Employment, Redundancy and unfair dismissal compensation changes.
3rd phase	Liquidation – employees' rights. Strikes – social security benefits. Lay-off pay Maternity pay and leave

Redundancy notification and consultation. (Handling redundancies).
Medical suspension.
Extension of Terms and Conditions.

2. Ensure one person in the Company is made responsible for: studying the Act in conjunction with this book and its particular implications for your company; obtaining further information on particular aspects from employers' organisations, etc; co-ordinating necessary administrative action (*see below*).

3. Make *administrative arrangements* to cover the following provisions in time for implementation at the statutory times:

Contracts of Employment *(see Chapter 3)*	Revise Statements for new and existing employees.
Part-time Employees *(see Chapter 3)*	Check Contract Statements of those working more than 16 hours or more than eight hours for five years.
Dismissal *(see Chapter 2)*	Guidance to managers on providing written statements of reasons for dismissal.
Pay Statements *(see Chapter 3)*	Prepare new format for written pay statements.
Safety Committees *(see Chapter 3)*	Arrange for nomination of safety representatives with recognised trade unions.
Time off *(see Chapter 1)*	Ensure procedures exist for: employees to claim time off from specified managers with authority to grant such permission; paid or unpaid leave as appropriate.
Redundancy *(see Chapter 2)*	Note that when employees are offered alternative jobs, a trial period is taken into account;

(see Chapter 7)	Set up administrative procedure for notification of redundancies to Department of Employment.

4. Revise existing, or prepare new, PROCEDURES and/or COLLECTIVE AGREEMENTS on the following (note that contractual arrangements which are better than statutory provisions must stand):

Lay-off *(see Chapter 1)*	
Suspension for Health and Safety reasons *(see Chapter 1)*	
Maternity Pay and Leave *(see Chapter 1)*	Especially arrangements for temporary cover, e.g. by part-time employees.
Redundancy *(see Chapter 7)*	Revise or prepare: procedure/agreement on consultation.
Shop Stewards' facilities *(see Chapter 1)*	Revise or prepare credentials covering time-off provisions.

5. Consider COMPANY POLICY and possibly issue guidelines and policy parameters to plants/subsidiary companies on the following:

Time off *(see Chapter 1)*	Consider whether or not to give *paid time* off for public duties and trade union activities. (Note that employers do not *have* to pay.)
Contracts of Employment *(see Chapter 3)*	Review all contractual terms (written or otherwise) to guard against breach of contract claims.
Dismissal *(see Chapter 2)*	Revise/prepare warning and disciplinary procedure.

Trade union membership and activities *(see Chapter 3)*	Review policies and procedures on recruitment, selection, promotion, etc., to ensure discrimination does not occur.
Recognition *(see Chapter 4)*	Review policy on recognition of trade unions especially in multi-union situations.
Information Disclosure *(see Chapter 4 and Appendix 3)*	Prepare positive policy on disclosure. Study ACAS Code when published, and discuss with unions.
Comparability Claims *(see Chapter 7)*	Consider plant and/or central negotiating structures. Such claims will, inevitably, be geared to low pay situations. Become an expert on your local labour market.

6. Finally, ensure company can defend itself well in industrial tribunal and other institutional proceedings by:

- having access to good legal advice
- maintaining clear written personnel records
- understanding the procedures used by industrial tribunals *(see Appendix 2)*.

Part I:
Individual Rights of Employees

Chapter 1: Time Off

There are now very few employees who are not given paid absence for annual and other holidays: paid holidays are now so much a part of life that it takes an effort to remember that before the war some manual workers got only a week's annual holiday and that was not always paid for. Holidays had to be saved for and were in fact sometimes referred to cynically as 'annual lock outs'.

We have come a long way since then; the Employment Protection Act takes us very much further. It deals with circumstances where the worker will be entitled to 'time off' because of circumstances not of his own choosing, for example, lay-offs because of shortage of work, suspension for certain medical reasons and the necessity to look for work when it is proposed that he be made redundant. The Act also deals with circumstances where time off is at the wish of the employee, such as time off for trade union activities, for public duties connected with local government membership or duties on the bench, and finally for maternity (though no doubt supporters of women's liberation would object to the classification of maternity as time off at the wish of the employee). And being human, employees will not only be keen to exercise new time-off rights, but to get paid for such time off work as well.

GUARANTEE PAYMENT FOR EMPLOYEES LAID OFF

We shall turn first to the guarantee payment sections of the Act. In practice this new right will generally be of importance only to hourly paid workers. It must be added however that in strict legal terms, it would appear in the light of cases such as *Jones v Harry Sherman* that it is not so straightforward to lay off manual workers (without a term in the contract) as has been commonly supposed. Salaried employees are not normally laid off (though they may give their consent to this) unless there is an express provision in their contracts of employment which gives management such a right. Guaranteed payments achieved by means of voluntary agreements are nothing new; such arrangements are also present in certain wages regulations orders made under the Wages Council Act. The agreements commonly provide for guaranteed weekly remuneration, i.e. the minimum amount of money payable, subject to certain conditions, whether the full number of hours have been worked or not. The payment of guaranteed weekly remuneration in these voluntary agreements and orders is made conditional on the worker being capable of and available for work and willing to perform such work as he could be reasonably expected to perform when work in his own occupation is not available. In some agreements the employer may give notice to terminate his obligation to pay, for example when work is not available owing to a strike or lock-out in the plant (in the case of the agreement between the Engineering Employers' Federation and the Confederation of Shipbuilding and Engineering Unions this exclusion extends to circumstances where work is not available because of a strike in *any* federated firm) or failure of, say, the electricity supply and so on.

What the Act does is to give every employee who has put in more than four weeks' service with his employer a *statutory* right to a guarantee payment which is calculated not on a weekly but on a *daily* basis. Thus three main categories of employees may be distinguished:

(a) salaried employees

(b) hourly paid employees who are presently covered by a voluntary guarantee payment agreement

(c) hourly paid employees who are not covered by any guarantee payment agreement.

The first and third categories are comparatively straightforward; the first will be entitled to the statutory guarantee payment prescribed by the Act but is not generally subject to lay-off; thus the new right, for this class of employee, is largely academic and of no practical significance. The third is entitled to the statutory payment and this entitlement will constitute a significant advance in his terms and conditions (and a consequent saving to the National Insurance and Social Security Funds). It is the second category – those employees who have dual entitlement – that presents certain problems. The employer may be regarded as being between the Scylla of breach of implied terms of the contract of employment and the Charybdis of breach of a statutory duty. He will have to compare the amounts (and qualifying conditions) of both the voluntary agreement and the statutory obligation and see that *both* are satisfied. This does not of course mean that he is obliged to pay twice – one obligation, if it is parallel, may be offset against another.

For example, a federated engineering employer who makes motor parts may find that he has to lay off workers because of a strike in British Leyland (which is also currently federated). The EEF agreement does not oblige him to pay guarantees; the statutory right given to employees by the Act does – so the employer must pay. If, however, the employer has already paid one week's guaranteed make-up under the Engineering Employers' Federation agreement in a particular quarter and that sum is greater than the statutory entitlement in that quarter then his statutory obligation to pay in the initial circumstances described falls away.

Despite these complications there is little doubt, however, that the main burden and expense of the guarantee pay provisions of the Act will fall upon small employers who have no voluntary agreement with their employees, and many of them will not have unions anyway.

When an employee (who must have been continuously employed for four weeks ending with the last *complete* week before what the Act refers to as the 'workless' day) is not provided with work on a day* when he would normally be required to work, because of a fall of activity or any other reason, then he is entitled to be *paid by his employer* a guarantee payment for that day or part of a day.

There are certain exceptions that should be noted. We have already seen that this right does not extend to an employee who has not completed four weeks' service; additionally, the Act provides that there is no entitlement to guarantee payment in circumstances where the employee is employed under a contract for a fixed term of 12 weeks or less, or employed to do a particular job which is not expected to last more than 12 weeks. Thus seasonal workers

*'Day' means any period of 24 hours from midnight to midnight; the Act takes note of two particular circumstances where a period of employment begun on any day extends over midnight into the following day, e.g.

(a) hours worked (or normally worked) before midnight are greater than those worked after midnight.

(b) hours worked (or normally worked) before midnight are less than those worked after midnight.

In case (a) the workless day is the first day; in case (b) it is the second day.

would generally not be entitled to guarantee payments, nor would, say, a house painter who is engaged to paint a house (though no doubt if the house was, for example, Blenheim Palace it could reasonably be argued that the painter was entitled to guarantee payment). Entitlement to payment would be extended, however, to contract employees who had been continuously employed by the same employer for more than 12 weeks prior to the commencement of the contract. Thus, for example, a builder could not absolve himself from the guarantee payment obligation merely by engaging brick-layers for successive contracts of less than 12 weeks (the 'lump' is another story).

Further, there is no entitlement to guarantee payment when work is not available because of a trade dispute involving the employer or an associated employer. It would seem that the definition of 'associated' will be the Trade Union and Labour Relations Act definition which employers should note is wider than that previously used in the Redundancy Payments Act.

Again there is no entitlement if 'alternative work which is suitable in all the circumstances' offered by the employer is 'unreasonably' refused by the employee. The wording of the phrase involving alternative work is carefully chosen; it implies that the alternative work offered by the employer need not be as closely defined as alternative work in a redundancy situation. It is perfectly possible to envisage a semi-skilled man being asked to do labouring work; that might be described as work suitable in the circumstances of potential lay-off. It is unlikely that an employer would get away with such an offer to a redundant semi-skilled man. Of course, in the guarantee payment situation the employee has the let out of saying that his refusal to do alternative work is not 'unreasonable'; no doubt there will be ample room for interpretation of that term in future industrial tribunal cases.

Finally, the Secretary of State is empowered to make an order excluding employees from the statutory entitlement to guarantee pay if they are covered by a collective agreement which may be substituted for the statutory provisions. An application must be made for exemption by all parties to the agreement and the agreement itself must satisfy stringent conditions.*

*The agreement must have procedures to enable an employee who claims he has not been paid a guarantee payment to pursue his claim; these procedures must include the right of access to arbitration or to an industrial tribunal.

The limit of the statutory guarantee payment is £6 a day (though the Secretary of State has power to apply for an increase in this, as in other, limits). The daily payments are also limited to the number of days an employee normally works – with a maximum of five – in each of the following periods in the year:

1st February – 30th April

1st May – 31st July

1st August – 31st October

1st November – 31st January.

These provisions imply that the maximum statutory entitlement per employee per year will be £120. It is of course perfectly possible for an employer to be obliged to pay sums in excess of that figure because of his obligation under a collective agreement; his payments under such an agreement would, provided all other conditions were equal (which as we have seen in the example of the strike-bound federated engineering company they may not be), be offset against, perhaps to the point of extinguishing, his statutory obligation.

The following formula is used to calculate the payment to be made on any given 'workless day':

(a) calculate 'guaranteed hourly rate'

$$\text{i.e.} \quad \frac{\text{amount of one week's pay}}{\text{number of normal working hours in a week}}$$

$$\text{(b) payment} = \frac{\text{no. of normal work hours in day} \times \text{guaranteed hourly rate}}{\text{no. of normal working hours in a week}}$$

Thus where one week's pay = £50
and normal week = 40 hours
and normal day = 8 hours

then guaranteed hourly rate $= £\frac{50}{40} = £1{\cdot}25$

theoretically guarantee payment $= £1{\cdot}25 \times 8 = £10$

but this is limited by the Act to a maximum of £6.

It should be noted that if short time working is in force when guarantee payment falls due to be paid, 'normal working hours' will be taken to mean normal working hours *before* the short time working started. Further, if the normal working week differs from week to week (or over a longer cycle) then 'number of normal working hours in a week' used in calculation (b) will be arrived at by the following formula:

(c) Number of normal working hours in week = $\dfrac{\text{Total number of hours over last 12 weeks*}}{12}$

Any employee who feels he should have received guarantee payment and did not, may complain, within three months, to an industrial tribunal. If the complaint is well founded, the tribunal will order the employer to pay the amount of guarantee payment found due by the tribunal.

These provisions of the Act will present employers with administrative complications and may, for employers not already contractually obligated by guarantee work agreements, be costly. They will undoubtedly save the State some money but, characteristically, at the expense of the employer.

PAYMENT FOR EMPLOYEES SUSPENDED FOR HEALTH OR SAFETY REASONS

The term 'medical suspension' appears, on examination of these provisions in the Act, to be ill-chosen. What the Act legislates for is not suspension on health grounds *per se* but on safety grounds. Fifteen enactments are listed by the Act, ranging from the Paints and Colours Manufacture Regulation 1907 through the Factories Act 1961 to the Radioactive Substances (Road Transport Workers) (Great Britain) Regulations 1970; it also lists the code of practice issued or approved under section 16 of the Health and Safety at Work Act of 1974. These all cover situations in which work of a particular kind – say the conveyance of radioactive substances by

*Ending with the last *complete* week before the workless day. Note that this calculation will not be affected by short time working which may be regarded as 'abnormal' hours.

road – can be stopped by statute until approved changes are made in equipment, protective clothing, method of handling, length of exposure, and so on. The worker literally cannot do the job he was employed to do; to allow him to do so would be in breach of one of the Acts (and the Code) specified.

There are some grounds for concluding that this could be a very important part of the Act; the section itself does not list the relevant statutes, regulations and codes but draws attention to those listed in the schedule relevant to the section. The Secretary of State may add to that list by order and it is understood that the TUC are already telling health and safety representatives to demand full pay for safety 'disruption' of this kind; thus we may assume that there will be pressure to extend the list in the schedule.

Now the problem arises of what the employer is to do with the employee until the changes are carried out. There are four possibilities and these are covered by the Act:

(a) the man may be dismissed
(b) he may be offered an alternative job
(c) he may be suspended on full pay
} in both cases until he can go back to his modified job
(d) he may have been made sick by doing the job and is incapable of working at all.

Dismissal. If an employee is dismissed in these circumstances, where it is a fair assumption that he should have been suspended on 'medical grounds', he will be able to take a case to an industrial tribunal for an unfair dismissal action and to do this if he has *four weeks*' continuous service and not, as in other circumstances, 26 weeks' continuous service. In such a case the man would not be entitled to a basic award (see Chapter 2 for new basis of compensation for unfair dismissal) but the tribunal will be able to make a compensatory award. This might very well be based on what he should have received on medical suspension.

If the employer decides to offer the employee *alternative employment,* the man or woman has the right to refuse such work if the refusal is not unreasonable. That means there may be situations in which it is reasonable for an employee to refuse what appears to be alternative work. Such offers of alternative work should be made in writing so that the employer may prove he has made such an offer and demonstrate to a tribunal that it was, in fact, reasonable.

If the employee is not dismissed and is not offered (nor unreasonably refuses) suitable alternative work, and he has four weeks' continuous service, he is entitled to be paid on *medical grounds* his full pay (i.e., not just his basic rate but all contractual elements of pay) for a maximum of 26 weeks. Any payment he is due under his contract of employment may be deducted from this 'suspended' pay. The assumption is that the changes required by the law in question will have been carried out in six months. The Act is silent as to situations where it takes longer, but an assumption might be that if he is dismissed after 26 weeks the man might be regarded as redundant.

If the employee is *unable to work because of sickness or injury* then he is *not* entitled to the medical suspension payment – a circumstance which emphasises the misleading nature of the term 'medical suspension'. A man is only entitled to the medical suspension payment if he is capable of work.

In certain circumstances, e.g., exposure to radiation, it may be possible for an employer to get someone else to do the job temporarily until such time as the original man is allowed by the relevant statute to return to the job in question. In these cases the temporary employee must be informed in writing that he will be dismissed when the suspended employee returns. It will not, however, be automatically fair to dismiss the temporary when the job holder returns. The employer will still have to show that the dismissal was reasonable in the circumstances.

Finally, the previous situation when a man applied for unemployment benefit and an insurance officer was prevented from adjudicating on his claim if there was an outstanding claim to a tribunal (which could be the case in guaranteed week cases as well as medical suspensions) has been changed. The insurance officer may now deal with claims for unemployment benefit on their merits – which will require his forming a judgement as to whether there is an entitlement to payment for medical suspension.

These provisions are novel in the employment field and the employer would be well advised to incorporate administrative procedures to deal with cases of medical suspension in conjunction with procedures he has devised to cope with the Health and Safety at Work Act. He should also be watchful for the trade union pressure to try to extend the lists of statutes, regulations and codes by industrial action. Such extension will have no legal force but may give rise to demands for a special collective agreement.

MATERNITY PAY AND LEAVE

The new rights which will be afforded to women employees in respect of pregnancy are likely to be regarded by many employers as radical, expensive and administratively difficult. Yet when one considers that these rights (or better) have been in existence for many years in other European countries without causing undue difficulties; that women have become a useful and substantial part of the labour force in recent years and that the demand from women for greater opportunity has become an accepted aspect of our society, these new legal protections relating to employment must be seen as both inevitable and necessary. It may be, however, that employers have had sound reasons for arguing that it is unreasonable that they, and not the State, should carry the cost burden. The women's liberation movement, however, has not been slow in observing that the right to return to work is but a legal nicety to a great number of women to whom lack of child care facilities will still be the greatest impediment to resuming work. For the time being, there is thus not likely to be a large increase in the number of women returning to work after childbirth, but employers can expect this position to change as greater pressure will undoubtedly be exercised by trade unions and women's movements on companies and the State to provide better crèche facilities. It is unlikely, in any case, that all the new rights in respect of maternity will be in force before 1977 partly because of the administrative difficulties of setting up the Maternity Pay Fund.

The new law affords protection to a pregnant employee in two ways. First, it makes it unfair to dismiss a woman employee solely on the grounds of pregnancy (except in special circumstances); secondly, it gives women the right to maternity pay and to come back to work within a specified period after confinement provided certain conditions are fulfilled. Before going on to detail these provisions, it is interesting to note that whilst the rights to maternity pay and return to work depend upon a two-year service qualification, there is no service qualification attached to the right not to be unfairly dismissed on grounds of pregnancy other than the six months which applies to unfair dismissal generally.

We shall now examine in detail:

(a) Pregnancy and dismissal.

(b) Eligibility for maternity pay and the right to return to work.

(c) Maternity pay, and the Maternity Pay Fund.

(d) The right to return to work.

Pregnancy and dismissal. It will be legally unfair to dismiss a woman (who has had at least six months' service) because of her pregnancy, except in the following circumstances:

(i) where her pregnancy has made her incapable of 'adequately doing the work' she was employed to do: *or*

(ii) where legal restrictions prevent her continuing her work whilst pregnant.

If, however, any suitable alternative vacancies exist in his organisation, the employer must first offer such jobs to the employee he proposes to dismiss for one of the two reasons specified above. Failure to do so will provide the employee with a right to unfair dismissal compensation. Suitable alternative employment must be employment which is suitable in relation to the pregnant woman in question and must be in a capacity and on terms and conditions not substantially less favourable than her old job. This definition of suitable alternative employment differs from that applied in the Redundancy Payments Act 1964 because here the terms and conditions, etc. can be less favourable though they must not be *substantially* so. Thus the range of alternative jobs which the employer must consider could be fairly wide.

Although not specifically required to do so by the Act, the employer would be well advised to put offers of alternative employment in writing as proof that a woman was given sufficient information to decide whether or nor the job was suitable.

Where no suitable vacancies exist, the employee can be fairly dismissed for one of the two reasons specified above, but if challenged in an industrial tribunal, the onus will be on the employer to show that no such vacancies existed. Employers should remember too, that the grounds of 'incapacity' are open to interpretation and can be challenged in industrial tribunal proceedings. For example, a tribunal may not necessarily accept that sickness absence amounts to 'incapacity'.

In effect, this section of the Act means that pregnant employees are free to decide when they wish to leave work. If they become incapable of doing their existing jobs, the employer is obliged to offer alternative employment as has been mentioned above. Obviously it will not be in the employee's interests to resign if, for example, her doctor declares her unfit to continue working, unless she has already reached the 11th week before the expected confinement. This is because if she resigns before this time she loses her right to maternity pay and to return to work after confinement. This assumes, however, that such women will be aware of their new legal rights and the fairly complicated legal provisions relating to such rights. This may not be the case, and employers may consider it more sensible in the long run to ensure that their female staff understand these procedures than risk dispute situations where women resign without realising that they may have forfeited rights by so doing.

The remedy for unfair dismissal is compensation awarded by an industrial tribunal (*see Chapter 2*). Such awards are calculated in basically the same manner as for unfair dismissal generally, except that:

> Compensation for an unfairly dismissed woman who would have had two years' service as at the 11th week before the date of confinement will take into account the loss sustained by losing her right to return to work (unless re-instatement and re-engagement is recommended). If, however, she is dismissed after the 11th week before confinement and during her absence, she does not lose her right to return to work. If she does return to work within the required time limit any compensation which was awarded to her in lieu of this right by the tribunal must be paid back to the employer if he so requests.

Eligibility for maternity pay and the right to return to work. Generally speaking, a pregnant woman with two years' service who has remained in employment up to the beginning of the 11th week before the expected week of confinement is entitled to six weeks' maternity pay and to return to work after confinement. There are, however, certain qualifications applying to these rights, as follows:

(i) Two years' service is calculated as at the beginning of the 11th week before expected confinement.

(ii) Remaining in employment until the 11th week before expected confinement means in effect that a contract of employment has remained until this date. Absence on account of sickness, for example, does not of itself affect the pregnant employee's rights.

(iii) A woman who has been *fairly dismissed before the 11th week* does not lose her rights whether or not she accepted an alternative job, if the reason for her dismissal was incapability or contravention of a Statute.

(iv) Both rights are conditional upon the woman producing a certificate indicating the expected date of confinement signed by a registered midwife or medical doctor, if requested to do so by her employer.

(v) The right to maternity pay is conditional upon the woman informing her employer (in writing if he so requests) that she will be absent because of pregnancy or confinement at least three weeks before her absence begins or 'as soon as is practicable'. The right to return to work is conditional upon her informing her employer that she intends to return to work in the same manner.

Maternity pay. The maternity pay which a woman will be legally entitled to receive from her employer (that is, women who meet the qualifications outlined above) is basically 9/10th of her week's pay up to a maximum of £80 per week for six weeks, minus social security benefit calculated at the flat rate national insurance maternity allowance. Earnings related supplements are thus not deductable from maternity pay and it is for this reason that the fraction 9/10th of a week's pay is used. Earnings related supplement is approximately 1/10th of weekly pay and therefore women receiving the full state maternity allowance will receive the equivalent of six weeks' full pay (up to a £80 maximum).

A 'week's pay' in relation to maternity pay is calculated in the same way as a week's pay under the Redundancy Payments Act 1964. In simple terms it means basic pay plus any other contractual pay obligations such as contractual overtime. The social security benefit which the employer may deduct is the basic State maternity allowance (£11·10 as from November 1975) and is deductible whether or not the woman concerned has been paying the full contribution rate and therefore whether or not she does in fact receive £11·10 from the State.

Example of maternity pay calculations. The maternity pay which an employer would have to pay an employee earning £90 a week would be calculated as follows:

Week's pay	£90·00	
Maximum allowed for calculation	£80·00	
9/10th of £80		£72·00
Deduct State maternity allowance		£11·10
One week's maternity pay=		£60·90
Six weeks' maternity pay=		£365·40

Maternity pay is payable for the first six weeks of absence falling after the 11th week before the date of expected confinement and could be paid either as a lump sum at the beginning of the period of absence or at the normal contractual payment times (i.e. weekly or monthly). It is essential to note that where an employer already has an established contractual arrangement for providing maternity pay, he does not have to pay both this and the new legal entitlement but must pay whichever is the better of the two entitlements as far as his female employees are concerned.

Maternity Pay Fund. The original intention of the Employment Protection Bill was that maternity pay should be borne and paid out directly by employers of women entitled to receive maternity payment. Due, however, to pressure from both sides of the 'House' and to representations from industry pointing out that this arrangement would be particularly onerous to employers of large numbers of women and might even have the effect of discouraging employment of women of child-bearing age, a last-minute amendment was introduced.

The Secretary of State will set up a Maternity Pay Fund – similar in operation to the Redundancy Fund. Employers will contribute to the Fund by way of a small increase in social security contributions (this will not, of course, apply to employees' social security contributions).

Employers will then be responsible for paying the statutory maternity pay directly to their employees but will be able to claim back the full amount from the Fund. Employers paying maternity pay which is part of a contractual obligation and in excess of the statutory amount, will still be able to claim back from the Fund that part of the payment which is equal to the statutory amount.

The Act allows that in certain circumstances, e.g., where the employer has been declared insolvent, the employee may be able to claim her statutory maternity pay directly from the Fund and the Secretary of State will have discretion as to whether or not to allow such a claim. Where the employer is not insolvent but the employee has been unable to extract payment from the employer and the Secretary of State is satisfied that (a) she is entitled to payment and (b) she has made every reasonable effort to persuade her employer to pay her (probably by seeking an industrial tribunal Order), then he may order that she be paid directly from the Fund and the employer's liability in respect of payment is transferred from the woman concerned to the Secretary of State and the Fund. Moreover, the Secretary of State will have power to refuse or reduce a subsequent rebate claim made by the employer if he is satisfied that refusal to pay represented default on the employer's part. In determining whether or not a woman's claim for direct payment from the Fund is reasonable, the Secretary of State may ask the employer to present him with relevant information, and refusal to do so may tender the employer liable to a fine of £100.

There will be a right of appeal to an industrial tribunal for:

(a) An employer where the Secretary of State has refused or reduced a rebate; *or*

(b) An employee where the Secretary of State has refused to accept her claim for direct payment from the Maternity Pay Fund.

Right to return to work. A woman who meets the required qualifications has a right to return to work at any time after confinement up to a maximum period of 29 weeks. She is entitled to return to work with her original employer and to the job in which she was employed under her Contract of Employment in existence before she left because of pregnancy or confinement. Where she accepted an alternative job because pregnancy made her incapable of carrying out her original job, it is the latter which is relevant to the return to the work situation and not the short term alternative job. The job to which she returns must also be on terms and conditions no less favourable than they would have been had she not been absent. This provision works both ways and imposes restrictions on both the employer and the woman concerned. It means that the woman has no right to return to a job which is suitable in relation to her new circumstances. For example, her original job might have involved shift work which, with a small baby, she would find it impractical to work. It is interesting to note the definition of 'job' in this context. This is taken to mean 'the nature of the work which she is employed to do in accordance with her contract and the capacity and place in which she is so employed . . . ' What this means is that it is the individual contract of employment which determines the job to which the woman is entitled to return. If, for example, her contract of employment defines her job as 'Secretary to the Sales Director' then there is clearly only one job to which she is entitled to return, but a contract which said 'A Director's Secretary' and either stated or implied that this applied to all directors' secretary posts would probably mean that the employer could quite legitimately offer her any of these posts on her return to work. Employers will be well advised to consider suitable wording in contracts of employment statements in the interests of allowing themselves at least a little flexibility.

What is of paramount significance here is the right of a woman to return to her job under the original contract of employment. If she is not permitted to return to work in such a job then she has a clear case of unfair dismissal against her employer. The only defence an employer has in such a case is that the original job has become redundant, in which case the woman is entitled to a redundancy payment. However, the matter is not quite so simple. If the original job has become redundant the employer must seek for and offer any suitable alternative work in his company or *in any of his associated companies*. If any such alternative jobs are available – and such a job may have less favourable terms and conditions provided they are not substantially so, and those jobs are suitable in relation to the woman – they must be offered. If such jobs exist but are not offered to the woman whose original job has become redundant, this amounts to unfair dismissal and *not* redundancy. It will clearly be sensible for an employer in such circumstances to offer a wide range of alternative jobs since he may not offer one which exists and which an industrial tribunal might subsequently consider to be suitable. Such job offers should be made in writing and should set out all the relevant terms and conditions so as to avoid misunderstandings in possible tribunal cases.

In order to exercise her right to return to work, a woman must notify her intended date of return to her employer not less than one week before she intends to return. The latest date she can return to work is 29 weeks after the date of her confinement with the following exceptions:

(a) She may delay her return by a maximum of four weeks beyond the 29th week if she notifies her employer not later than the expiry of the 29th week that she is medically unfit and produces a medical certificate to this effect. She may only exercise this right of extension once.

(b) The employer, having been notified of the woman's intended date of return may delay that time for a maximum of four weeks, provided that he notifies the woman, specifying the reasons for the delay and giving a new date for her return to work.

(c) If the woman has notified a day of return but it happens that on that day there is an interruption of work due to industrial action or for some other reason, she is entitled to delay her return until work has resumed or as soon as reasonably practical thereafter. If she has not notified a day but because of an interruption of work it is not practical to expect her to return before the expiry of the 29th week, then she is entitled to return when work has resumed or not later than 14 days thereafter.

The woman concerned is entitled to return to the job she held under the original contract of employment on terms and conditions not less favourable than if she had not been absent. In effect this means that any improvement in pay or conditions which occurs during the period of her absence must be applied to her on her return. Her period of absence does not break her continuity of employment but the period of absence itself is not counted as a period of employment for the purpose of calculating benefits. Thus an employer is not obliged to make up pension contributions for this period of absence, but on return to work she does not lose pension seniority accrued *at the time she left*. However, it should be noted that the period of absence must be counted in calculating length of service for notice rights and redundancy payment. Since, moreover, the period of absence does not break continuity of employment, the two years' service qualification before a woman is entitled to maternity pay and to return to work may already have been met when she returns to work and she would, theoretically, again be entitled to time off for pregnancy immediately. It would be an unfortunate employer who allowed a woman to exercise her right to return to work only to find that she was six months pregnant again!

The Government's original Bill gave some help to employers faced with the difficulties of keeping jobs open for women absent because of pregnancy and entitled to return to work. This was by way of allowing that a replacement in a job left vacant by a pregnant employee could be fairly dismissed on the woman's return to work provided that on engagement, the position was made clear in writing. Unfortunately the Act as now drafted is somewhat less helpful since it is no longer automatically fair to dismiss an employee in the above circumstances. The employer must also have acted 'reasonably' in the circumstances. The effect of this is not entirely clear and the position may only be clarified by case law established in the industrial tribunals. It is likely, however, that the employer will at least have to consider the temporary replacement for alternative jobs which might be vacant. Employers may wish to consider employing temporary replacements on a part-time basis (i.e., less than 16 hours per week) or on contracts for less than six months' service since in both cases, the employee has no right to claim unfair dismissal.

MATERNITY – SUMMARY OF ENTITLEMENTS

EXAMPLE	ENTITLEMENTS
A. Leaving Work Because of Pregnancy	
A.1. Employee resigns* before 11th week before expected confinement.	No entitlement under the Act.
A.2. Employee resigns any time after 11th week before expected confinement	
– if she has 2 or more years' service	MATERNITY PAY; RIGHT TO RETURN TO WORK.
– if she has less than 2 years' service	No entitlements under Act.
A.3. Employee with 2 years' service fairly dismissed (i.e. on grounds of incapability or contravention of a Statute) and	RIGHT TO MATERNITY PAY AND RETURN TO WORK. [NOTE that employee can dispute the grounds of 'capability' and thus claim unfair dismissal in which case see A.5 below.]
– not offered suitable alternative vacancy *which exists*	May claim unfair dismissal. If tribunal recommends re-engagement or reinstatement, will also be entitled to maternity pay and right to return to work.
– refuses suitable alternative vacancy	RIGHT TO MATERNITY PAY AND RETURN TO WORK but cannot claim unfair dismissal.
– no suitable vacancy exists	MATERNITY PAY AND RETURN TO WORK.

*An employee who resigns may decide to claim constructive dismissal in which case she has access to the same legal entitlements as a pregnant employee claiming unfair dismissal.

MATERNITY – SUMMARY OF ENTITLEMENTS

EXAMPLE	ENTITLEMENTS
A.4. Employee with less than 2 years' service fairly dismissed	No entitlement under the Act but see NOTE under A.3 above.
A.5. Employee unfairly dismissed (i.e. for reasons other than those specified in A.3 above) and has:	
– 2 or more years' service.	RIGHT TO CLAIM UNFAIR DISMISSAL. If dismissal found to be unfair, can be awarded compensation and/or reinstatement or re-engagement. If the latter awarded, employee regains right to MATERNITY PAY and RETURN TO WORK.
– less than 2 years' service but at least 6 months' service.	Right to claim UNFAIR DISMISSAL.
A.6. Employee fairly dismissed and then resigns during her period of notice.	Loses entitlement to claim unfair dismissal because employer had no opportunity to offer her alternative job.
A.7. Employee with 2 years' service unfairly dismissed during *her absence* after 11th week before expected confinement.	Retains RIGHT TO RETURN TO WORK, but if she does so, after having been awarded unfair dismissal compensation, her employer has the right to claim back from her the compensation so paid.

MATERNITY – SUMMARY OF ENTITLEMENTS

EXAMPLE	ENTITLEMENTS
B. Returning to Work after Confinement	
B.1. Employee gives one week's notice of intention to return to work not more than 29 weeks after confinement [leave may be extended for further 4 weeks if medical certificate proving unfitness produced to Employer].	Entitled to return to her JOB* (on not less favourable terms than if she had not left). Employer may delay return for 4 weeks for specified reasons. * Specified in her contract before she left.
B.2. Employee notifies intention to return to work as above but original job has become redundant and:	
– no suitable vacancies exist in Employer's or associated company	entitled to REDUNDANCY PAYMENT.
– offered suitable alternative vacancy	Loses entitlements under the Act if she refuses but the job must not be substantially less favourable.
– not offered suitable alternative vacancies where these can be shown to exist in the employer's company or associated companies.	Entitled to claim UNFAIR DISMISSAL compensation.
B.3. Employee notifies intention to return to work as in B.1 above, but employer refuses to reinstate her for any reason other than redundancy.	Entitled to claim UNFAIR DISMISSAL compensation.

All examples given apply only to women entitled to return to work because they had appropriate service qualifications, etc. See A above.

TRADE UNION DUTIES AND ACTIVITIES

For the first time this Act will give shop stewards* or, as appropriate, union members in companies a statutory right to time off – paid in relation to 'trade union duties' and paid or unpaid for 'trade union activities'. Neither of these rights will, however, come into force until ACAS has produced a Code of Practice providing 'practical guidance' on the amount of time off which is reasonable and the purposes for which time off should reasonably be provided. This will be no easy task for ACAS since practice varies so widely from industry to industry and indeed within industries.

At one end of the spectrum is the factory where one or more shop stewards are paid by the company but attend to their union duties on a full-time basis. In such a factory, the new law is unlikely to affect the general situation. At the other end of the spectrum is the company where the shop steward normally carries out his trade union duties during lunch hours or after working hours. The latter case must be exceedingly rare these days. Indeed the role and status of the shop steward generally has become increasingly important in many industries with the growth of shop floor activity and militancy. Theoretically the amount of time off which a shop steward requires should be determined by the volume of trade union business with which he has to deal at the workplace, but some employers will argue that Parkinson's Law applies and that the more time he has at his disposal the more problems the shop steward will find to plague the employer. There may be a grain of truth in such an argument, but on the other hand the interests of good industrial relations at the workplace are not always best served if the shop steward does not have sufficient time off for adequate consultation with the union members he represents; to conduct joint business with management and for proper contact with the officers of his union. The new legal provisions of the Act together with the expected ACAS guidelines may make a real contribution towards better workplace industrial relations if they stimulate management and unions to reach practical and clear agreements on the role of shop stewards and the facilities they are provided with.

*In fact the Act refers to 'trade union official' in connection with these rights but for the sake of simplicity, the term 'shop steward' or 'trade union representative' is used in this chapter.

We shall now examine in detail the two separate provisions relating to time off for trade union representatives and members.

Time off for trade union duties. An employer must allow his employees who are representatives of an independent trade union recognised* by him, paid time off for carrying out those of their duties as representatives which are concerned with industrial relations matters between the employer (or an associated employer) and his employees. In addition, the employer must also allow representatives paid time off 'to undergo training in aspects of industrial relations relevant to the carrying out of those duties'. ACAS is under specific statutory instructions to produce a Code of Practice giving guidance on what is reasonable time off and 'the purposes for which and the occasions on which' such time off may be taken. Moreover, ACAS must include special reference to the circumstances under which trade union representatives may take time off for matters concerned with industrial action. The drafting of the Act on this particular point is interesting for it implies that there may be circumstances under which an employer will be required to give paid time off to shop stewards involved with industrial action. Clearly there will be cases where this will be very much in the employer's interest – for example where the shop steward is trying to dissuade some of his workers from a lightning and unofficial walk-out. On the other hand, it is to be hoped that ACAS will not suggest that an employer should pay a shop steward whilst he is initiating such a walk-out!

The Act refers to 'trade union official' in connection with the right to paid time off. The most important trade union official in this context is of course the shop steward but as shall be seen from the definition which follows, other officials such as a Branch Secretary are also included and may therefore also be entitled to these new rights.

'Trade union official' is defined by the Trade Union and Labour Relations Act 1974 as 'any person who is an officer of the union or of a branch or section of the union *and who (not being such an officer) is a person elected or appointed in accordance with the rules of the union to be a representative of its members or of some of them, including any person so elected or appointed who is an employee of the same employer as the members or one or more of the members, whom he is to represent.*' It is the latter part of this definition (in author's

*'Recognised' means recognised by the employer for purposes of collective bargaining.

italics) which applies to time off rights. According to the law, therefore, it may be entirely up to the trade union concerned to decide how many 'officials' (or shop stewards) there shall be in any particular establishment who will then be entitled to these new legal rights. The present Code of Industrial Relations Practice does, however, recommend that management and unions should reach joint agreement on the number of shop stewards and the 'work groups for which each steward is responsible' and no doubt the specialised ACAS Code on this matter will make a similar recommendation. In any event it will now make sense for employers who do not already have such agreements to attempt to get them before this part of the Act comes into operation.

The right to paid time off applies only to duties which the shop steward carries out in accordance with his function as shop steward and connected with joint business between the employer and his employees. Presumably, therefore, a shop steward representing union members in the tool room will not have the right to take time off to represent members in the maintenance section (unless the matter concerns the tool room members in some way). Similarly he will not have the right to paid time off for the purpose of carrying out his duties as the union's branch secretary although as we shall see, he may be entitled to reasonable time off although not necessarily with pay, to be involved in purely trade union activities. As drafted in the Act, however, the interpretation of 'duties' is by no means clear and could be subject to wide interpretation. One can foresee that there will be arguments on the shop floor relating not simply to whether or not time off should be granted to a shop steward but as to the particular industrial relations business with the employer which he is entitled to attend. In addition to studying the ACAS Code carefully, an employer would be well advised to set up some kind of joint procedure for considering such claims (if in dispute) before they become full blown disputes or are referred to industrial tribunals.

The pay to which a shop steward is entitled for these periods of time off must be calculated on the basis of what he would have earned had he not been a shop steward (where his remuneration as an employee varies with the amount of work done). Thus pay must be calculated according to his average hourly earnings or 'if no fair estimate can be made of these earnings, the average hourly earnings of work of that description of persons in comparable employment with the same employer or, if there are no such persons, a figure of average hourly earnings which is reasonable in the circumstances'. Employers of manual workers with complicated wages structures will be well advised to reach agreement with the trade unions on the method of calculation to be used. Where, however, such an agreement already exists it should be noted that any pay a shop steward receives under such an agreement may be offset against the legal liability to pay.

Time off for trade union activities. Time off for trade union *activities* differs from time off for trade union *duties* described above in three basic respects:

(i) The employer need not pay the employee for time off under this section;

(ii) The right to time off for trade union duties applies to any or all employees who are members of a trade union recognised by the employer;

(iii) Trade union activities are not confined to industrial relations matters between the union and the employer as is the case with trade union duties. 'Activities' in this section of the Act means any activities of an appropriate trade union (i.e. an independent union recognised by the employer) or any activities in relation to which the employee is acting as a representative of such a union.

The only specific qualification which the Act applies to the definition of 'trade union activities' is that such activities should not themselves consist of industrial action. Thus an unofficial meeting called for the express purpose of disrupting the work does not attract permitted time off within the meaning of the Act – although in factories where this is already an established form of industrial action, it will no doubt continue unaffected by the new law. It is arguable, however, that since time off for such purposes will clearly be unauthorised both by the employer and by the law, that tribunals may well find dismissal as a result of such action fair. As with time off for trade union duties, ACAS is under a statutory duty to produce a Code of Practice giving guidance as to the amount of time off for trade union activities and the purposes for which time off is given which is 'reasonable in all the circumstances'. ACAS must also give specific guidance on 'whether and the circumstances in which a trade union member is entitled to take time off under this section for trade union activities connected with industrial action'.

Once the ACAS Codes appear, the new rights for time off for trade union representatives and members will come into force and employers will be well advised to have agreed procedures on time off covering not only the circumstances justifying time off (in the light of ACAS guidelines) but also specifying from whom permission may be sought, and an appeal procedure to a higher level of management in the event of disagreement.

Remedies. Any employee who is a member of an independent and recognised trade union may complain to an industrial tribunal that his employer has:

(a) unreasonably refused to permit him time off for trade union duties (if he is a trade union official)

and/or

(b) refused to pay him time off for trade union duties

or

(c) unreasonably refused to permit him time off for trade union activities.

The tribunal, if it finds the complaint justified, may make a declaration to that effect and/or make an award of compensation. Such compensation may be assessed not only in relation to any loss occasioned to the employee but also on consideration of the employer's 'default'. An application to a tribunal must be presented within three months of the complaint occurring unless this is not reasonably practical.

The circumstances which would be considered reasonable justification for refusing time off for trade union duties and activities will only become clear as a result of future industrial tribunal case law and by reason of the forthcoming ACAS guidelines. However, it is difficult to imagine a tribunal finding it unreasonable to refuse time off in the case of a two-man business or if the time off would result in having to close down a plant or business and thus cause serious financial damage.

PUBLIC DUTIES

Employees will have the right to time off (not with pay) for carrying out the following specified public duties:

(a) a justice of the peace

(b) a member of a local authority

(c) a member of any statutory tribunal

(d) a member of, in England and Wales, a Regional Health Authority or Area Health Authority, or, in Scotland, a Health Board

(e) a member of, in England and Wales, the managing or governing body of an educational authority, or, in Scotland, a school or college council or the governing body of a central institution or a college of education; or

(f) a member of, in England and Wales, a Water Authority or, in Scotland, river purification board.

Reasonable time off must be permitted for employees who are members of any of these bodies to attend meetings of the body and to attend to activities approved by the body and relating to the carrying out of its functions. Employers are not required to pay employees for such time off from work, although some employers do, as a matter of personnel policy, provide full pay for certain specified public duties minus any fees received by the employee from the body to which he is giving his service. Such policies will not be affected by the Act.

There are two notable exclusions from the Act's list of public duties which qualify for time off. Voluntary Reserve Service with the armed forces, which most people would include under the general heading of 'Public Duty' is not included and the reason for this exclusion is not clear. Jury service is not included simply because attendance for jury service is already compulsory by law except in special circumstances.

Unlike the provisions relating to time off for trade union duties and activities, ACAS is not required to produce practical guidance in the form of a Code of Practice on what is reasonable in relation to time off for public duties. The Act, instead, lists three factors which should be taken into account by the employer in considering how much and on what occasions he should permit time off. These are:

(a) the amount of time required for the duty in question

(b) how much time off the employee in question has already had for trade union duties and/or activities; and

(c) *the circumstances of the employer's business and the effect of the employee's absence on the running of that business.*

The first factor (a) above will be difficult for employers to assess unless they have a detailed knowledge of the particular public body in question. The third factor listed above will perhaps have most relevance to employers, particularly to small employers. A man running a two assistant shop, for example, might properly contest that he could not possibly allow time off for one of those assistants to attend to Town Council business. In the longer term, industrial tribunal rulings will help to clarify what is 'reasonable' in relation to time off for public duties.

Remedy of a breach of the right to time off for public duties is to complain to an industrial tribunal who may award compensation on the same lines as in contended cases of time off for trade union duties and activities.

LOOKING FOR NEW WORK WHEN REDUNDANT

The last of the new rights to time off introduced by the Act is for 'reasonable' time off, with pay, for workers under redundancy notice, for the purpose of seeking new employment or to make arrangements for training for future employment.

All employees who will have had two years' continuous service at the expiry of their redundancy notice (or at the expiry of the minimum period of notice to which they would be entitled under the Contracts of Employment Act 1972) are entitled to this new right. The Act does not require ACAS to produce a Code giving guidance on what is 'reasonable' in relation to the amount of time off required. Effectively, however, the Act qualifies 'reasonable' by limiting the amount of pay which an employer can be required to give an employee to two-fifths of a week's pay. There may, however, be situations in which the employer can reasonably refuse any time off – for example if the employee is a key member of a team or in a very small business where he literally cannot be spared.

Pay in respect of absence under this section of the Act is calculated at the 'appropriate' hourly rate.

Where pay does not vary with the amount of work done, this means:

$$\frac{\text{One week's pay}}{\text{Normal working hours per week}} = \text{Hourly rate}$$

Where pay varies from week to week, it means:

$$\text{Average working hours per week} = \frac{\text{Total number of hours worked in last 12 weeks before notice given}}{12}$$

$$\frac{\text{One week's pay}}{\text{Average working hours}} = \text{Hourly rate}$$

Remedy. An employee may complain to an industrial tribunal that he has not been allowed reasonable time off, or that his employer has not paid him for time taken off work in accordance with his entitlement under the relevant section of the Act. Complaints must be made within three months. The tribunal may make a declaration on the issue and/or order the employer to pay the employee the amount of pay due to him. If the employee was not allowed to take time off, and the tribunal finds that the employer acted unreasonably, then it may order the employer to pay the employee up to two days' pay. This is the maximum which may be awarded by the tribunal by way of compensation.

In practice, this means that employers will have to allow employees serving out redundancy notice two days off with pay for the purpose of seeking alternative work or making training arrangements. If this is impractical, the employer may wish to give the employee, with his agreement, two days' pay in lieu of such leave. In any event, many employers already provide such facilities for time off during redundancy notice and, at least in some cases, more generous facilities.

It is difficult to see how employers will be able to check that the employee is actually using the time off for the purposes defined by the Act and for this reason it seems likely that, in practice, the two days' paid leave during redundancy notice will become an automatic right.

NEW TIME OFF PROVISIONS*

ITEM	PURPOSE AND DESCRIPTION	CRITERIA FOR ELIGIBILITY	MAXIMUM AMOUNT OF TIME OFF	PAID OR UNPAID	REMEDY	OTHER NOTES
Guarantee payment	Lay-off pay	4 weeks' continuous service (fixed term contracts of 12 or less weeks excluded) Has not refused suitable alternative work	5 days per quarter	Paid – maximum daily payment of £6	Complaint to industrial tribunal – order for sum due	Not payable if lay off is due to trade dispute in employer's (or associated company's) premises
Medical Suspension	Employee not allowed to do his normal work because of health or safety hazard	4 weeks' continuous service; Has not refused suitable alternative work	26 weeks on full pay	Paid	Complaint to industrial tribunal – order for sum due	If employee dismissed, may claim unfair dismissal with only 4 (not 26) weeks' service
Maternity	Pay on leaving because of pregnancy and right to return to work after confinement	2 years' continuous service and remaining in employment up to 11th week before expected confinement	11 weeks before expected week of confinement plus 29 weeks after confinement (possible extension to 33 weeks)	6 weeks' pay ($\frac{9}{10}$ of a week's pay maximum £80) minus flat rate national insurance payments	Complaint to industrial tribunal – order for the sum due and for compensation for unfair dismissal	May claim unfair dismissal if dismissed because of pregnancy (only 6 months' service required) or if denied right to return to work

Trade Union Duties	Time off for employees who are trade union officials to attend to industrial relations matters connected with the employer and his employees and for training	Must be a trade union official as defined by the TULRA	'Reasonable' and with reference to ACAS code (not available at time of writing)	Paid – full pay – or where pay varies, his average hourly earnings or those of similar employees	Complaint to industrial tribunal – declaration as to rights and/or compensation	
Trade Union Activities	Time off for employees who are trade union members to take part in their unions' activities	Must be member of an independent trade union which is recognised by his employer	'Reasonable' and with reference to ACAS code	Unpaid	Complaint to industrial tribunal – declaration as to rights and/or compensation	Entitlement does not apply to activities which amount to industrial action
Public Duties	Time off for various specified public duties such as J.P.	Membership of specified body	Unspecified – but in relation to amount of time required, time off already taken and circumstances of employer's business	Unpaid	Complaint to industrial tribunal – declaration as to rights and/or compensation	
Redundancy	To seek new employment or make training arrangements under redundancy notice	2 years' continuous service	'Reasonable' but see 'Remedy'	Paid – normal pay	Complaint to industrial tribunal – declaration and/or order payment (maximum of 2 days' pay)	

*Generally speaking, where there are contractual arrangements in respect of any of these items, e.g. a guaranteed week agreement then the employer must meet the better of the two obligations.

Chapter 2:

Dismissal and Redundancy

DISMISSAL

The then novel concept of unfair dismissal was introduced by the Industrial Relations Act 1971. Employees who fulfilled the various qualifications were given the right not to be unfairly dismissed. In the years following the passing of the Industrial Relations Act a large body of experience has been built up by employees, by trade unions, by employers' organisations and most importantly by industrial tribunals, in the application of the law on unfair dismissals. Some changes have also been made since 1971, notably the reduction of the qualifying period for claims from two years to six months and the introduction, by the Trade Union and Labour Relations Act 1974, of the concept of constructive dismissal (i.e., where an employee terminates his contract in circumstances in which his employer's conduct towards him forces him to resign and amounts to dismissal).

Further changes introduced by the Trade Union and Labour Relations Act on dismissals were:

(a) abolition of the provisions enabling the employer who dismissed a man by reason of trade union pressure upon him to conjoin the union as a third party in any subsequent unfair dismissal action

(b) extension of the time limit for complaints from 28 days from the effective date of termination to three months

(c) explicit extension of the burden of proof to the employer, not only to prove that an allegedly unfair dismissal was in fact justified, but also to proving that in the circumstances he acted reasonably

(d) the ability of tribunals to recommend (but not order) re-engagement was extended to enable them also to recommend reinstatement

(e) the upper limit of financial compensation was raised from £4,160 (i.e., 104 weeks at £40) to £5,200

(f) in consequence of the repealing of the right to belong or not to belong to a trade union, it was provided that it was unfair to dismiss an employee for being, or proposing to be, a member of an independent trade union (or for refusing to be, or become a member of a union which is not independent).

Thus one major concept, unfair dismissal, established by the Conservatives in 1971, far from being swept away by the wind of political change, was added to and extended by Labour's Trade Union and Labour Relations Act. The Employment Protection Act carries this extension, as we shall see, a good deal further.

Written statement of reasons for dismissal. Employers have generally been somewhat reluctant to give a dismissed employee written reasons for dismissal. Such caution is understandable; quite apart from the possibility of a libel suit, such a statement would certainly be used as evidence in any subsequent unfair dismissal case, and reasons written down in the heat of the moment might not reflect too well upon an employer. The Act will change this situation. It gives an employee who has been dismissed (in precise terms been given notice of termination of his contract of employment, or if his contract of employment has been terminated by his employer without notice, or if he is working under a fixed term contract which is not renewed), the right to be given, upon request, a written statement which will give particulars of the reason for dismissal. This right will be enjoyed only by employees who have been, or will have been, continuously employed for 26 weeks ending with the *last complete week* before the effective date of termination. It should be noted that this statement *will* be admissable for evidence in any subsequent proceedings such as suits for wrongful dismissal, application for redundancy, cases of alleged unfair dismissal and perhaps even in some instances, libel suits.

In the light of these provisions an employer might well wonder what will happen if he does not comply with such a request within the prescribed period of 14 days from the date of the request. In short, the employee refused such a written statement will be able to present a complaint to an industrial tribunal (which complaint should in normal circumstances be presented within three months of the refusal) which shall, if the complaint is held to be justified, make an award equal to two weeks' pay. Further, if the employee *is* given a written statement which he considers is inadequate or untrue he may also complain to a tribunal which may make a declaration as to what it finds the employer's reasons were for dismissal, and also make an award of two weeks' pay. In any case the failure of an employer to provide a written statement, or his providing an inadequate or false statement later corrected by a tribunal would clearly do him little good in a subsequent – and one would think, inevitable, unfair dismissal case.

Orders for reinstatement or re-engagement. When the Industrial Relations Act was enacted, one of the criticisms of that Act by trade unionists and their supporters was that it did not provide for automatic reinstatement, nor re-engagement. True, a tribunal could recommend re-engagement and if this recommendation was ignored by the employer then the tribunal could increase the compensation award. The Employment Protection Act provides that when a tribunal finds that a complaint of unfair dismissal proved, then the tribunal shall be obliged to explain to the dismissed employee that the law now allows, in certain circumstances, for orders for reinstatement or re-engagement to be made. Such orders will only be made if the complainant wishes the tribunal to make such an order. There may occur cases in which a dismissed man would not, in reality, wish to return to the firm which dismissed him unfairly (and paid compensation to him) but no doubt employees will be advised in many cases to ask the tribunal for an order for reinstatement or re-engagement.

A reinstatement order means that an employer has to treat the employee he dismissed unfairly as if he had not been dismissed. The order will specify:

(a) monies payable to the employee in respect of arrears of pay (including bonuses) for the period between the *date of termination* and the *date of reinstatement*. In calculating these arrears of pay, the tribunal has to reduce the amount by wages in lieu of notice, or ex gratia payments paid by the employer, pay from other employment during the period of absence, social security benefits which the employee may have received and such other benefit as the tribunal thinks appropriate in the circumstances. It is also likely that these amounts will be taken into account by the tribunal in assessing the amount of the compensatory award to which the dismissed employee will be entitled

(b) any rights and privileges (such as seniority and pension rights) must be restored to the employee

(c) the date by which the order must be complied with

(d) that he should also receive any improvement in wages and/or conditions which have been made during the period of his absence.

A re-engagement order means that the employer (or the successor of the employer or an associated employer) must give the dismissed employee a job comparable to that job which he lost when he was unfairly dismissed or a job which constitutes 'suitable employment'. The re-engagement order will state:

(a) identity of the employer (which can be important in cases where the company has changed hands or has changed name)

(b) the nature of the employment

(c) the remuneration for the employment. In addition the re-engagement order must, like the reinstatement order, state:

(d) any money payable to the employee in respect of arrears of pay for the period between the date of termination and the date of re-engagement

(e) rights and privileges to be restored to the employee

(f) the date by which the order must be complied with.

The sequence of decisions for a tribunal is thus first, decide if the complaint of unfair dismissal is well founded; second, tell the complainant of the possibility of reinstatement and re-engagement orders; if the complainant wishes such an order the tribunal must, third, consider the possibility of reinstatement as a first priority. In so considering this possibility, the tribunal is obliged to take into account the wishes of the complainant regarding reinstatement and whether it is *practicable* for the employer to reinstate the person. It should be noted that this consideration of practicability will *not* take into account the fact that an employer has engaged a permanent replacement for the dismissed person *unless* the employer can show that he could not get the work done without so engaging a *permanent* replacement or that he engaged the replacement after what the Act describes as a reasonable period without having heard from the dismissed employee that he wishes to be reinstated or re-engaged. How the employer is supposed to get the work done while waiting the 'reasonable period' to hear from the dismissed employee is another matter.

In cases where the dismissed man contributed to some extent to the dismissal the tribunal must decide whether it would be just to order a reinstatement. Fourthly the tribunal, if it decides against reinstatement, must then consider re-engagement, again taking into account the wishes of the complainant (who may prefer re-engagement to reinstatement because he will not have to work for the particular person who dismissed him), whether it is practicable for the employer (and in the case of re-engagement the successor of the employer or an associated employer) to comply with a re-engagement order, and again if the dismissed man contributed to some extent to the dismissal whether it would be just to order re-engagement.

Thus the tribunal has a more complex task than ever in an unfair dismissal case. And the employer has whole new burdens laid upon him.

Compensation for unfair dismissal. The Employment Protection Act states that a tribunal will compensate for unfair dismissal by firstly a *basic award* and secondly a *compensatory award.* In addition, if an order for reinstatement or re-engagement is not fully complied with, the tribunal may, upon application, award further compensation for failure to comply with the order.

The **basic award** is calculated thus:

(a) Work out the period, ending with the effective date of termination during which the employee has been continuously employed

(b) work out the number of years of employment falling within that period and allow

(i) $1\frac{1}{2}$ weeks' pay for each year which *wholly consists* of weeks in which the employee was not below the age of 41.

(ii) 1 week's pay for each year which consists wholly of weeks in which the employee was below the age of 41 and was not below 22.

(iii) $\frac{1}{2}$ week's pay for each year which consists wholly of weeks in which the employee was below the age of 22 and was not below 18.

(c) The calculation above is reduced by 1/12th for each month of service between 64 and 65 for men and 59 and 60 for women.

(d) There will be a basic award of two weeks' pay where the tribunal finds that the dismissal constituted redundancy and that the employee

(i) was not entitled to a redundancy payment (e.g., did not have the requisite service)

(ii) is not treated as dismissed for the purposes of the Redundancy Payments Act

(iii) is entitled under the Redundancy Payments Act to less than two weeks' pay (which in fact means he gets the two weeks' pay)

(e) The maximum amount of a week's pay is to be taken as £80 per week

(f) If the tribunal finds that the dismissed employee caused or contributed to the dismissal it shall reduce the amount of the basic award by such proportion as it considers just and equitable

(g) The amount of the basic award will be reduced by the amount of any redundancy payment awarded in respect of the same dismissal (which in practice would be calculated by (a) to (e) above).

The **compensatory award** is calculated thus:

(a) an amount of money considered by the tribunal to be just and equitable in relation to the loss (legal expenses not included) sustained by the dismissed employee by reason of his dismissal. In this latter connection it should be noted that though the loss in question must be attributable to action taken by the employer, it is not likely that loss would be assessed in the circumstances of a dismissed employee in a blind rage assaulting his wife and being fined £20. In trying to assess how far the loss was attributable to action of the employer, the tribunal is not permitted to take into account industrial action exercised on the employer to get him to dismiss the employee. Thus if a union pressurises an employer to, say, get rid of a non-union employee, any loss attributable would be the same as if the employer had been a free agent

(b) the loss will include any expenses reasonably incurred by the dismissed man in consequence of his dismissal with the exception of legal expenses as previously noted. Such losses would include though, the cost of looking for another job

(c) the loss will include any benefit which the dismissed employee might be reasonably expected to have had but for the dismissal (though in the case of monies obtained by redundancy – either statutorily or voluntarily – the benefit by which that benefit would have exceeded a *basic* award)

(d) the dismissed employee will have a duty to mitigate his loss

(e) where the tribunal finds that the dismissal was occasioned by any action of the employee it shall reduce the amount of the compensatory award by such proportion as it considers just and equitable

(f) the amount of the compensatory award is limited to 104 weeks pay or £5,200, whichever is the lower.

Penalties for non-compliance with a reinstatement or re-engagement order. Somewhat analagous, at least in operation to the sections on compensation for unfair dismissal, are the sections of the Act dealing with compensation for non-compliance with an order for reinstatement or re-engagement. There are two cases to be considered. Where the employer has in fact reinstated a dismissed employee, or as the case may be, re-engaged him, in accordance with an order of a tribunal, but has failed to comply with all of the terms of the order. In this case a tribunal shall make an award of compensation of such amount as the tribunal thinks fit having regard to the loss sustained by the employee. This compensation – which, bear in mind, goes to a man who is back in employment with his old employer – is, like the compensatory award described above, limited to 104 weeks' pay or £5,200. The second case concerns the situation where either a reinstatement or re-engagement order is not complied with at all by the employer. Whilst the term 'order' is used it is plainly not possible to regard such an order in the same way as, say, a High Court order; in the same way that the law cannot be used to make someone do a specific job so it is inconceivable to use the law to force an employer to take on (in this case reinstate or re-engage) a particular man for a certain task. Thus in cases of non-compliance a tribunal will compensate for unfair dismissal with a basic and compensatory award and in respect of the employer's failure to comply with the order to reinstate or re-engage, it will award further compensation to the dismissed employee. The only defence the employer may use in this circumstance is that it was *not practicable** to comply with the reinstatement or re-engagement order. If the dismissal in respect of which the order had not been complied with is either:

(a) unfair because of membership or non-membership of a union or for taking part in the activities of an (independent) trade union

(b) a dismissal which is an unlawful act of discrimination under the Race Relations Act; or

(c) a dismissal which is unlawful under the Sex Discrimination Act

then the tribunal may award up to 52 weeks' pay but not less than 26 weeks' pay.

*It should be noted that if the employee has expressed a wish to be reinstated or re-engaged with his ex-employer at least seven days before the tribunal hearing, that regulations for the tribunals will provide that the employer will be required to pay the costs or expenses of any postponement of the hearing caused by his (the employer's) failure to put forward reasonable evidence as to the *availability* of the job from which the man was dismissed, or of a comparable job.

If the dismissal does *not* fall under one of these three headings (one is enough – the compensation does not triple up in the case of a black female shop steward whose dismissal may contain a bit of all three), then the tribunal may, in the event of non-compliance with a reinstatement or re-engagement order award up to 26 weeks' but not less than 13 weeks' pay to the dismissed man or woman. Thus the 26–13 weeks' range will constitute the norm; the 52–26 weeks' range the heinous 'triple' offences.

The following example illustrates the seriousness of the sums involved:

Case study – dismissals. Man aged 45 (birthday 20th April, 1930) earning £70 a week who has worked for his employer since 25th February, 1955 is dismissed on 18th August, 1975 for, say, *misconduct.* He is a union member who is trying to get the union recognised and contests that his dismissal was related to this activity. The tribunal finds his dismissal unfair, explains the reinstatement and re-engagement orders to him; he says he wishes to be reinstated and the employer refuses to have him back. Possible compensation awarded might be:

(a) basic award: 16 × 1 week's pay = 16 × £70 = £1,120
$\quad$ 5 × 1½ " " = 5 × £105 = £525

(b) compensatory award, say, = £250
(loss of wages, net of tax, for 10 weeks less social security payments received).

(c) compensation for non-reinstatement
52 × £70 = £3,640

£5,535

For his loss the employee receives the sum of £5,535, all of it payable by the employer (who may also have incurred legal costs). Had the same man been made redundant he would have received £1,645 of which the employer would have received a rebate of £823.

These sums serve to emphasise the care which employers should use in considering dismissals. If we add the continuance of possible industrial pressure upon employers because of e.g., 'give Mr. X his job back or we strike' to these new obligations the employer's lot is clearly not a happy one.

Also the employee need not wait until his contract terminates before putting in his claim for unfair dismissal as he can start proceedings the moment he is given notice. For employees with longer periods of notice this makes reinstatement, or rather withdrawal of notice, most valuable.

It is no wonder that the TUC is trying to step up the pressure to get laws which will limit self-employment. Such are the burdens added to the conventional employment relationship by the procedures and penalties outlined above that it is possible that employees (for whom, self-employment offers many short term advantages) will collude even more with employers to work on self-employment contracts.

Entitlement to the right not to be unfairly dismissed. Eligibility to the right not to be unfairly dismissed (and access to the newly enhanced scale of awards) is opened up considerably by the Act.

We noted that the Trade Union and Labour Relations Act had brought down the qualifying service from two years to one year (and now six months). The Trade Union and Labour Relations Act also extended the application period to a tribunal from 28 days to three months. Now, the Employment Protection Act reduces, in the case of part-timers, from 21 hours per week to 16 hours the minimum number of hours worked per week required to qualify for protection for unfair dismissal. (In addition someone who works as little as eight hours per week will qualify with five years' service at that rate.) It extends to *all* employees the right previously only enjoyed by those in employment with employers who had four or more employees. This will bring many farm workers and shop assistants, as well as domestic servants, into the scope of the dismissal provisions. And the Act no longer excludes those in employment who are close relations of the employer (except if the relationship is that of spouse – a husband may fire his wife, or vice versa, without fear of ending up in an industrial tribunal as well as, possibly, the divorce court).

Special provisions with regard to dismissal connected with trade union membership or activities. When the Government introduced these provisions into the Act, the Minister of State said that they were prompted by a Department of Employment analysis (covering the period 1966–73) of strikes which resulted from dismissal. This analysis showed that there was a dramatic increase over the period in man days lost by industrial action resulting from the dismissal of workers' representatives. The Minister claimed that the new provisions were designed to offer a statutory remedy as an alternative to industrial action in this type of case. One would earnestly hope so, for they involve radical and unusual features which many employers will regard as the thin edge of the wedge leading to a situation where they may not dismiss until a tribunal gives them leave.

In short the Act provides that a dismissed employee may present a complaint to a tribunal* that the reason, or principal reason, for his dismissal was that he was, or proposed to become, a member of an independent trade union, or that the reason was because of his actual, or intended activities (e.g., collecting subscriptions, taking part in meetings and so on) in the union. A tribunal will consider complaints of this type only within a time limit of up to seven days immediately following the effective date of termination (i.e., when the contract is broken and money paid in lieu, or at the expiry of a notice period). There is no provision allowing an extension of that time period.

Also within this time limit, the tribunal must receive a written certificate from an 'authorised' official (which can mean either a full time officer of a union or a shop steward) of the dismissed man's union (or intended union) stating that *he* considers there are reasonable grounds for supposing the dismissal is connected with actual or intended union membership, activities, etc.

As soon as practicable after receiving the application and the official's certificate, a tribunal is required to give the employer at least seven days' notice of the hearing (as well as copies of the application and certificate). The determination of the application by the tribunal amounts to an interim hearing to establish if there is a *prima facie* case that the employee has been dismissed on grounds of trade union membership or activities.

If the tribunal finds there are grounds for believing that the complaint is justified it will ask the employer if, pending the full hearing of the case, he will reinstate, or if not, re-engage the employee. If the employer is willing to do so the tribunal will issue an order. If the employer fails to attend the hearing, or is willing neither to reinstate nor re-engage, then the tribunal will make an order for the renewal or continuation of the man's contract of employment. This really

*Which may, for this interim relief purpose, consist only of a Chairman.

means that the employee is, in effect, suspended on full pay. The choice for the employer is either give him back his old job, with all losses made up, or give him a suitable alternative job, or pay him for doing nothing. In point of fact the device of suspension is often used in a case of this sort and there have been some notable cases where men have been suspended for months to prevent industrial action being taken. Despite the new procedure there is of course no guarantee that industrial action will not be taken, and indeed one can envisage situations in which the employer will have to face the legal sanctions described here and a strike as well.

In cases where the employer does not comply with the order for the renewal or continuation of his contract, the dismissed employee may complain again to a tribunal. The tribunal will determine the amount due to him probably at the same time as it carries out the final hearing of the unfair dismissal case. If the case is upheld then the tribunal will award a basic award, a compensatory award and the sum due for non-compliance with the 'suspension' order. If the non-compliance with the interim order is in respect of reinstatement or re-engagement then the tribunal will of course apply the compensation formula described on page 58 ff.

These new dismissal provisions represent a formidable package for employers to cope with. Simple redundancy will seem more and more a calm and cheap haven to which to repair. Redundancy itself however, has been altered by the ubiquitous hand of the Employment Protection Act and we now turn to those changes.

REDUNDANCY*

The Act makes it clear that if an employer makes an employee an offer, whether in writing or not, before the ending of his employment under the previous contract, either to renew his contract of employment, or to re-engage him under a new contract, then under specific conditions, if the employee unreasonably refuses such an offer, he will not be entitled to a redundancy payment.

The conditions are that the new contract takes effect immediately on the ending of his employment or after an interval of not more than four weeks thereafter; that in either case the new job would not

*The Act's provisions relating to consultation with trade unions during a redundancy situation are outlined in Chapter 7.

differ in respect of the ability of the employee to do it, or of the place where it is to be done, or of its terms and conditions, from the old job (or if different would represent an offer of 'suitable' employment). Such offers need not be in writing.

Trial period. Prior to the enactment of the Employment Protection Act, where the contract under which an employee worked was terminated by the employer and a new contract on different terms was offered in writing, the employee could not claim redundancy pay if he worked under the new contract before deciding he could not really live with the new job and resigned. Tribunal decisions established that he could be entitled to a so-called 'trial period' if he stated he was only trying the job out. The amendments introduced by the Employment Protection Act have the effect of formalising this trial period to a minimum of four weeks* in which the employee may work at the new job (which as we have seen may be covered by the same contract, or one which constitutes suitable employment) without losing his right to a redundancy payment. In other words, simply by taking the new job he does not put himself in a position where he has burned his boats as far as getting redundancy pay is concerned.

If, however, the employee *unreasonably* terminates the contract during the trial period he is not entitled to a redundancy payment in respect of his old job. But the Act makes it clear that if the employee terminates the contract during the trial period for a valid reason (i.e. not unreasonably) he retains his redundancy rights. For that matter, if the employer, for a reason connected with the change to the new contract (i.e. the man is clearly not up to it) terminates the contract during the trial period *then* the employee is held to have been dismissed on the date when his original contract expired.

Thus there are the following possibilities to be considered:

(a) the employee is dismissed and no new offer made: he may, *inter alia*, claim redundancy

(b) the employee is dismissed (i.e. his old contract is terminated) and he is offered, within four weeks, a new contract for the same job or a new contract for a suitable alternative job;

*There will be cases where a longer trial period may be agreed upon for the purpose of retraining the employee under the new contract, which in this case would not of course be the same contract. Such an agreement must be in writing and made before the employee starts the new job.

if he refuses such an offer, he cannot claim redundancy (though he may dispute, of course, the suitability of the alternative job); if he accepts it he can still terminate the new contract and claim redundancy with the date of his dismissal being the date of termination of the old contract

(c) the employee loses his right to a redundancy payment if he unreasonably terminates the new contract during the trial period

(d) If the employee completes the trial period of four weeks (which may be extended by the offer once again of a new contract or a 'suitable contract' or indeed may have been extended beyond four weeks by prior agreement) then he loses any claim to redundancy in respect of his old contract. He does not, however, lose continuity of service, and if made redundant at some future date in the life of the new contract will calculate his service from the beginning of the old contract.

These provisions establishing a trial period have been generally accepted as reasonable and they should not cause undue difficulty to employers. They concentrate, sensibly, not upon ensuring that all employees get a redundancy payment, but upon maintenance of employment and continuity of employment.

Delayed claims for redundancy. Under the 1965 Redundancy Payments Act an employee could only make a claim for redundancy pay within six months of being made redundant. There have been some hard cases which have involved delay occasioned by the difficulty of an employee establishing which employer he had to make a claim to in circumstances where his original employer had, say, merged with another, and that other with another, and so on. By the time he figures out who should be the recipient of the claim (which may involve a failed claim to the wrong employer) he could run out of time.

The Employment Protection Act now provides for a second six month period following the first on the following conditions:

That the employee *does* enter a claim which can be either –

(a) *de facto*, by agreeing and receiving payment, or

(b) by making a claim in writing to the employer, or

(c) by appealing to a tribunal in respect of redundancy, or

(d) (and this is added by the new Act) he has made a complaint to a tribunal that he has been unfairly dismissed in respect of his termination *and* shows good reason to a tribunal that his failure to make a claim by any of (a)–(c) should be set aside in the interest of justice and equity.

This extension to a second term of six months will complicate life for the employer, though it would be unreasonable to criticise the provisions on that score. It does mean, however, that the employer will not know for a year whether an ex-employee proposes to claim redundancy and this will cause administrative and costing problems. It should be noted that if the employer voluntarily pays a redundancy payment during the second six months he will be entitled to a rebate from the Redundancy Fund provided he can satisfy the Secretary of State that it would be just and equitable to do so (which means the employer can show he had nothing to do with the delay).

Other provisions relating to redundancy. Other minor amendments made by the Act are as follows. *First*, as in the Trade Union and Labour Relations Act, an employee will be regarded as being dismissed by his employer if he terminates his contract with, or without notice, by reason of the employer's conduct. This provision introduces the concept of 'constructive' dismissal and allows for a situation where a man may be patently redundant but is forced to sit at an empty desk all day in somewhat embarrassing circumstances. His resignation in these circumstances will not preclude him from receiving a redundancy payment.

Secondly, the Act again follows the Trade Union and Labour Relations Act by amending the definition of an associated employer to cover the case where the association is between companies controlled by a third person. This has the effect of an employee's continuity of service being preserved if he moves not only to an associated employer (or that associated employer takes over his contract) but also moves to a company which is not associated in the previous definition with his old company, and is in fact associated with another company which has both his old and his new company in association – though not one with the other. This provision again improves the maintenance of employment and is generally welcomed. *Thirdly*, the Act confirms the limit of £80 for calculation of redundancy payment and imposes a duty on the Secretary of State to review this limit. There is also provision in the Act for a substantial increase in the statutory borrowing provisions applied to the Redundancy Fund, a measure which reflects not only the increased limits but the increased possibility of successful applicants which the changes described here imply.

None of these changes are such as to invalidate the view that employers will prefer, at least in individual cases, redundancy claims to unfair dismissal claims. Another problem, already manifesting itself in these days of economic crisis and straitened company finances, is that some small employers literally cannot afford to make members of their employees redundant because their liquidity situation does not permit payment of redundancy pay. The Act is silent as to that sad situation, which can really only be solved by an increase in employer contributions to the Redundancy Fund and an increase in rebate, a change which would not be welcome to some differently placed employers. If the present economic crisis deepens however, we may possibly anticipate yet another amendment on some such lines to add to the list described in this section.

SAMPLE DISCIPLINARY PROCEDURE

(Agreed between.........(the Company) and.........(the Union(s))*
on.........(date).

PREAMBLE:

This agreement applies to...
........................(Description of employees concerned).

The procedure below must be followed in all cases where an employee's performance or conduct is unsatisfactory and could lead to a written warning, suspension or dismissal – subject to the exceptions mentioned in this agreement (refer to written disciplinary rules where these exist).

PROCEDURE:

STEP I

The employee's immediate supervisor has a talk with him pointing out the deficiencies, suggesting methods of improvement and specifying a time limit† for such improvement. In such cases, additional training may be appropriate and should be considered. If, at the end of the time limit, no improvement has taken place, then:

STEP II

The employee's immediate supervisor gives a formal oral warning to the employee specifying his deficiencies, giving a plan for improvement and a time limit for its accomplishment. The employee must:

- have been informed of his right to be accompanied by a Union representative or workplace colleague, whilst the warning is being given;
- be allowed to state his case.

The supervisor must make a written summary of the conversation and send a copy to Personnel/Administration Department.

At the expiry of the stated time limit, the supervisor must review the situation with (the appropriate Personnel Officer). If a marked improvement has occurred, then a note to that effect should be sent to the employee (copy to Personnel Department). If not, then:

STEP III

A further meeting should be arranged with the employee (accompanied by a Union representative or workplace colleague, if he so desires), his manager and (the appropriate Personnel Officer). At this meeting, a written warning will be given to the employee specifying:

- a clear statement of the deficiencies;
- a plan and time limit for improvement;
- that redeployment, demotion or dismissal (as the case may be) could result if there is no improvement within the specified time limit, or if there is a reoccurence of the misconduct.

*If relevant.

†The time limit will vary according to the gravity of the performance deficiency or misconduct. During the time set, the individual must be regularly informed of his progress or otherwise.

The employee should sign a copy of this letter which should be sent to Personnel Department.

The situation must be reviewed at the expiry of the time limit. If satisfactory improvement has resulted, the employee must receive a note to this effect (copy to Personnel Department). If not, then:

STEP IV

Before redeployment, demotion or dismissal is effected, Personnel/Administration Department must be consulted to ensure that the procedure has been complied with. When the action to be taken has been decided, the employee is to be given appropriate notice (in writing) in an interview with his manager, who should send a written note of the proceedings to Personnel/Administration Department. The appropriate Union representative should also be informed.

STEP V

Appeal

An individual may lodge an appeal within 24 hours of any action taken under Step IV above to his (senior manager). Such appeals will normally be considered within 5 working days. His employment will not be terminated within this time limit if an appeal has been lodged. If a further meeting/interview is arranged, the individual has the right to be accompanied.

Gross Misconduct

In cases of gross misconduct (e.g. assault) the procedure above may be by-passed and the individual may be immediately suspended without pay with a warning that dismissal will probably follow shortly, but not before 72 hours have elapsed. An appeal may be lodged within this time to the (senior manager). Cases of gross misconduct justifying suspension and instant dismissal are rare.*

Serious Misconduct

In cases of serious misconduct, Steps I and II above may be by-passed and the individual warned (in Step III) that an occurrence of the misconduct may/will lead to dismissal.

Suspension from Duty

Circumstances† may arise which lead the Company to suspend an individual from duty. Such suspension may be with or without pay.

Action against a Union Representative

The Industrial Relations Code of Practice requires at Paragraph 133 (v) that 'no disciplinary action should be taken against a shop steward until the circumstances of the cases have been discussed with a full-time official of the Union concerned'.

*The Employer may also find it useful:

(a) to specify those managers who have the authority to take such action in a separate brief for managers; and

(b) to require that Personnel/Administration Department be consulted, in any case, before such action is taken.

†For example, where an Accounts Clerk is suspected of 'fiddling the books'.

Chapter 3: Miscellaneous

MORE NEW RIGHTS FOR WORKERS AND TRADE UNIONS

The Employment Protection Act represents something of a gamble by Parliament that many of the causes of friction in industrial relations can be reduced, or even eliminated altogether, by giving legal backing to perennial demands made of employers by the trade unions. The theory is that the more that areas of traditional contention can be narrowed, the likelier it is that the unions will settle down to a more responsible role within the community. Psychologically, the analysis has been made that the trade union movement, because of its history and its background, suffers from an aggressive inferiority complex. If the cure requires employers to be disadvantaged in a legal sense, then so be it: they, after all, will be the main beneficiaries of a more peaceful and less strife-torn system of labour relations. Indeed, many employers will be much more worried about the additional costs that will fall on them, not only in terms of potential penalties and legal costs, but also in administrative overheads, than they will be about the scales of industrial tribunal justice being weighted against them.

We have already seen that important concessions have been made by the Act to the welfare of employees, to their extra-mural existence on local authorities and on the magisterial bench, to the working of their trade unions. We have also seen a general toughening up of their rights not to be unfairly dismissed, and their right to compensation if they are made redundant. But the Act contains a number of other important extensions of the rights of unions and employees: sufficient, indeed, for this measure to have been awarded the soubriquet of the 'Trade Union Benefit Act'.

In this chapter we deal with half a dozen areas that come under the general heading of improvements in the rights of employees. They cover the priority given by the Act to the employee as a creditor of an insolvent company, changes in the Contracts of Employment Act, new regulations on pay statements, the right of an employee not to be discriminated against on account of his trade union activities, an important modification of the Health and Safety at Work Act, and an extension of the jurisdiction of industrial tribunals to hear certain cases covering breaches of contracts of employment. All are important.

INSOLVENCY – RIGHTS OF EMPLOYEES IN RESPECT OF UNPAID WAGES, ETC.

Getting a legal judgement against a company which goes bankrupt is a sad experience. In general terms you line up with the other creditors strictly ordered according to the law and if the well dries up before you get to it then hard luck. Unpaid wages fare rather better than debts since the Bankruptcy Act of 1964 and the Companies Act of 1948 specify that unpaid wages are paid in priority to all other debts an insolvent employer has incurred – even before the claims of debenture holders.

The section on insolvency in the Employment Protection Act puts the following payments in the same bracket as unpaid wages:

see Chapter 1
- (a) guarantee payments
- (b) payment for suspension on medical grounds
- (c) maternity pay
- (d) payment for time off in connection with carrying out trade union duties and for payment for time off to look for work

see Chapter 7
- (e) payment in respect of a protective award granted in connection with failure to consult a trade union in redundancy situations

However, the Act goes further than simply giving priority for payment of these new benefits which it brings into being. On application by an employee (or employees) the Secretary of State may pay certain debts of the insolvent employer out of the Redundancy Fund (i.e. debts in respect of (a)–(e) above, as well as arrears of pay unpaid in lieu of notice, holiday pay and unfair dismissal compensation limited to £80 in respect of one week and limited to an aggregate of eight weeks.)

To do this the Secretary of State is obliged to satisfy himself that the employer has become insolvent and that he owed his employees money under the various provisions on the date of his insolvency (or on the date of termination of the employee's contract if that occurred after the insolvency). The exception to this 'relevant date' provision is in respect of maternity pay, in which case the relevant date is the date of the woman's application to the Secretary of State *(see Chapter 1)*. In the same circumstances of insolvency, employer contributions to an occupational pension fund which were unpaid at the time of the insolvency may also be paid from the Redundancy Fund.

In all the situations where money (subject to the specified limits) is paid by the Secretary of State out of the Redundancy Fund, payment would not be made until a statement of the amount due had been secured from the trustees in bankruptcy or the liquidator (though if such a statement is delayed beyond six months the Secretary of State may pay anyway).

Conversely, the Act allows an applicant who has asked the Secretary of State to pay such monies out of the Redundancy Fund and has been told that the payment is due may complain to a tribunal (within three months of the Secretary of State's decision) that the money has not been paid, or some but not all has been paid. Again, those who have applied to the Secretary of State for payment of occupational pension scheme monies may appeal to a tribunal that the money has not been paid.

Where payments have been made from the Redundancy Fund, the rights and remedies of employees (and occupational pension funds) will be transferred to the Secretary of State and any monies recovered by the Secretary of State under those rights and remedies will be used to reimburse the Redundancy Fund.

The idea behind these provisions is simple enough. It is that the new rights conferred upon employees by the Employment Protection Act are not going to be allowed to be fruitless because an employer becomes insolvent. No problem arises for the beneficiaries if the monies can be recovered on a priority creditor basis: if not, the sum due will be paid out of the Redundancy Fund, and if there is an eventual shortfall then the Fund will have to stand it. One might deduce that in a period of uncertain trade and a high level of bankruptcies that the drain on the Fund – even on a cash flow basis – would be considerable and might be given as a reason for increasing employer contributions.

So anxious is the Act to see that due money is received that, as we have seen, it allows the Secretary of State to be subject to a declaration of a tribunal if he is negligent or laggard in paying the money by taking it out of the Redundancy Fund.

Finally, the Secretary of State will be able to require the production of information and documents from employers in connection with applications under these provisions. Penalties of a fine not exceeding £100 for refusal or wilful neglect, and a fine of up to £400 for knowingly or recklessly making false statements can be imposed.

CONTRACTS OF EMPLOYMENT ACT 1972 – AMENDMENTS

The main effect of the Contracts of Employment Act which was introduced in 1963, was to affect the employment contract in the following two ways; it required that certain specified terms of the contract must be set out in a written statement and given to the employee not later than thirteen weeks after the start of employment; and it laid down statutory periods of notice which an Employer was required to give an employee before terminating his employment. The Industrial Relations Act of 1971 amended this Act first by increasing the minimum periods of notice to: one week after thirteen weeks of employment; six weeks after ten years' and eight weeks after fifteen years' service; and also by requiring that written statements should include details of grievance procedures and a statement of rights in connection with trade union membership. The Trade Union and Labour Relations Act 1974 which overtook the Industrial Relations Act, retained these rights with the exception of the rights relating to trade union membership.

The Employment Protection Act continues the general theme by once again amending the Contracts of Employment Act to provide longer periods of notice and adding to the length of the statutory statement of terms and conditions of employment. The main changes to the Act are designed to:

(a) Increase the statutory minimum periods of notice;

(b) Require that details of disciplinary rules and job titles be included in the statement;

(c) Require that previous service with another employer, which counts as continuous service, be specified in the statement; and to

(d) Amend the list of categories of employees excluded from the Contracts of Employment Act's general provisions: by changing the definition of part-time employment and continuous service, and narrowing the definition of a close relative.

Minimum periods of notice which an employer is required to give an employee if he wishes to terminate his contract are extended to one week for each year of service up to a maximum of 12 weeks after 12 years' service. These are, of course, statutory minima and where the employer and employee have agreed on longer periods of notice, then these will continue to apply. Indeed there have been cases where the contract was silent as to notice and the courts have declared a 'reasonable' notice period which is longer than the statutory minimum.

Contract of employment statements (which must now be issued to the employee after four weeks' continuous service) must now specify any disciplinary rules which are applicable to the employee or refer to a document which sets out any such rules. This does not mean that where there are no disciplinary rules, these must suddenly be invented. It is possible, however, that industrial tribunals will not look kindly upon employers accused of unfair dismissal, who claim a breach of disciplinary rules which were not referred to in the employee's contract of employment statement. It is also necessary to specify in the statement the person to whom the employee may appeal if he is dissatisfied with a disciplinary decision, and the 'manner in which any such applications should be made'. If there is, in addition, a more elaborate appeals procedure, then this must be outlined in the statement itself, or the statement must refer the employee to a document which outlines this procedure. The statements must now, therefore, cover details not only of grievance procedures but also of disciplinary rules and appeals procedures. However, there is one specific exception. The statement is not required to cover grievances or rules or disciplinary procedures relating to health and safety. Presumably, this is because it is intended that statutory regulations relating to Health and Safety Committees will cover these aspects. The statement must also state the employee's job title and this will mean that if his job title subsequently changes for some reason, this change to his original statement must be notified to him in writing.

It will thus be important for employers to ensure that job titles and subsequent changes are accurately stated or they may be in difficulties if faced with claims of unfair (possibly constructive) dismissal or breaches of contract relating to a change in job.

Continuous service. It will also be necessary, where the employee has had previous service with another associated employer, which counts as continuous service, to put this in the statement, mentioning also the date upon which such previous service commenced. In addition, where there has been a change of employer, or in the name of the employer, the written statement informing the employee of such a change must state the date upon which service with the previous employer started (assuming that such a change does not, within the meaning of the Act, break continuity of service).

Excluded categories. There are now two basic changes to the list of categories of employees who are excluded from the provisions of the Contracts of Employment Act and who are therefore not entitled to minimum periods of notice or to written statements of their terms and conditions.

First, close relations of the employer who are excluded from the Act will now be restricted to 'husband and wife only'. Thus a son of an employer who runs a family business will have the protection of the Contracts of Employment Act if he works for his father, as well as protection from unfair dismissal. Secondly, the definition of part-time employees is changed. Under the Contracts of Employment Act 1972, employees who worked for less than 21 hours a week were excluded from the Act's provisions. The position now will be that only those who work less than 16 hours a week will be excluded – but an employee who works for less than 16 hours but more than eight, for a period of five years will then be afforded the protection of the Contracts of Employment Act, and his service over the preceding five years will count as continuous. A further amendment to the definition of continuous employment relating to service with an associated employer must also be watched since such service must be counted in calculating the employee's notice entitlement.

ITEMISED PAY STATEMENTS

The Truck Acts of the 19th and early 20th century form the background for the new legal provisions dealing with pay statements. The Truck Acts were introduced primarily to stop the abuses of the 'truck system' whereby employers paid partly in money and partly in kind so as to compel their employees to buy goods at the employer's own shop. Thus, the Truck Acts made payment by any means other than in the 'current coin of the Realm' illegal. The Payment of Wages Act 1960 modified these Acts by allowing employers to pay wages by cheque, money order or postal order, provided that the employees agreed to the arrangement. This same Act also required employers to provide itemised pay statements setting out the gross and net amount of wages together with a description of any deductions made from gross wages. However, the Truck Acts and the Payment of Wages Act applied only to manual employees and their employers. The Employment Protection Act's requirements for itemised pay statements apply to *all employees* and furthermore allow complaints of non-compliance to be made to the industrial tribunals.

In practice, most employers today provide employees with an itemised pay statement – particularly where a computer pay-roll system is in use – and many of such existing statements will meet the new legal requirements. In any event, it is hard to dispute the logic of letting employees know exactly how the pounds in their pockets at the end of the week are made up and why the total amount differs from the gross wages stated in their contract of employment statements.

Under the new Act, every employee will be entitled to receive from his employer an itemised pay statement at or before the time his wages are paid. Such a statement must specify:

– the gross amount of wages or salary;

– the net amount of wages or salary;

- the amount and purpose of any fixed or variable deductions (fixed deductions do not need to be separately mentioned if covered by a standing statement – see below);
- the amount and method of payment if different parts of the net payment are paid in different ways.

Fixed deductions. The required statements which must accompany the payment of wages may simply state an aggregate amount of fixed deductions provided that the employee in question is given a standing statement of fixed deductions and that such statements are updated and reissued at least every twelve months. Standing statements must specify:

- the amount of each deduction
- the intervals at which deductions are made
- the purposes for which they are made.

Any changes to the deductions mentioned in standing statements must be notified to the employee in writing before the change is effected. Provided this is done, it is not necessary to issue a new standing statement every time a change is made but, of course, the whole statement must be updated at twelve-monthly intervals.

Generally speaking, most deductions will not be 'fixed' where the employee's pay varies from week to week depending on the amount of work done (e.g. piece workers). Thus National Insurance contributions will only be 'fixed' deductions where pay does not vary. Income tax is unlikely to be a fixed deduction even where pay does not vary since tax varies not only with income but with changes in personal circumstances.

Gross wages and methods of payment. There is no requirement to itemise the various elements which make up gross pay – e.g. overtime, shift allowances and basic rates. (Although the Contracts of Employment Act 1972 requires such itemisation in the written statement of terms and conditions.) This seems a rather strange omission since a bonus payment can hardly have much effect as an incentive if the employee does not know the amount he has earned. However, the Act does give the Secretary of State power to add to the list of particulars which must be itemised in pay statements so this omission may well be rectified at a later stage. In any event, many employers already state these amounts in existing pay statements if only because separate itemisation is necessary for accounting purposes.

Computerised pay-roll systems may, however, have to be re-programmed to allow for itemisation of all deductions, and it may not be easy to allow for all possible permutations – for example, deductions for maintenance payments made under a Court Order.

The requirement to state the amount and method of payment harks back to the Truck Acts and subsequent Payment of Wages Act 1960 mentioned above. Where part of pay is made by money or postal order and part in cash, then each amount must be set out separately against the method of payment.

Remedy. An employee may present a complaint to an industrial tribunal that his employer has not provided him with an itemised pay statement as required by the Act; or that the statement provided to him by his employer does not contain all the particulars required by the Act. If the employee is making such a complaint against an employer whose services he has left, then the complaints must be made within three months of the date his employment ceased. The industrial tribunal will be empowered to make an order in respect of the missing statement or, as the case may be, the missing particulars. Furthermore, if deductions have been made from the employee's salary which have not been specified in his pay statement, then the tribunal will be able to order the employer to pay the employee an amount equal to all the unnotified deductions over a period of 13 weeks immediately prior to the date the employee made his application to the tribunal.

TRADE UNION MEMBERSHIP AND ACTIVITIES

Section 5(1) of the Industrial Relations Act 1971 provided that every worker was to have the right to belong to any registered trade union of his choice, the right to take part in the activities of that union at the appropriate time, and the right to seek and accept appointment or election to office in his union.

The Industrial Relations Act also gave workers the further right not to be dismissed, penalised or *otherwise discriminated against* by reason of his exercising his right under Section 5. That fell with the other repealed parts of the Industrial Relations Act by reason of the passage of the Trade Union and Labour Relations Act 1974. However Section 24 of the old Act, which provided, *inter alia*, that it would be regarded as unfair dismissal if the reason (or principal reason) for a dismissal was the employee in question exercising his rights under Section 5(1). was preserved by Schedule 6(4) of the Trade Union and Labour Relations Act. Thus, by reason of that Act, it was still regarded as unfair dismissal if a man was dismissed because of trade union membership or activitities, provided the union was independent.

The Trade Union and Labour Relations Act did not, however, try to go further, as did the Industrial Relations Act, and cover the situation where a man is *discriminated against* by his employer by actions which stop short of dismissal. Thus actions constituting 'penalising', or 'otherwise discriminating against', an employee by an employer for trade union membership or activities are not mentioned by the 1974 Act and did not, before the passage of the Employment Protection Act, allow for the possibility of an action on these grounds by an aggrieved employee. The Employment Protection Act attempts to plug this gap by providing that an employer who prevents or deters (or who refuses to employ) an employee from exercising his rights in respect of independent trade union membership may be the subject of a complaint by the employee to an industrial tribunal.

In specific terms the Act protects an employee from action by his employer which has the effect of:

- preventing or deterring him from taking part in the activities of an independent trade union at any appropriate time or penalising him for doing so: or

- compelling him to be or to become a member of a trade union which is not independent.

These are clear enough once we have taken aboard the further definition of 'appropriate time' (*see below*), but as we shall see, it is significant that there is now no mention, as there was in the Industrial Relations Act of 'discriminating against'.

Deterrent action of the extreme sort represented by dismissal is already covered by the Trade Union and Labour Relations Act. However, we might well ask what kind of action the Employment Protection Act has in mind when it describes 'action (short of dismissal)'. We would suggest that there are two different classes of such action.

(a) actions which involve an individual employee as such. Examples of such action might be:

- transferring the employee to a less agreeable (but not sufficiently different as to invite a redundancy claim) job because of trade union activities, e.g. recruiting new members
- not selecting an employee for overtime
- allocating him a 'dud' machine

and so on

(b) action which involves an employee as a trade union member. An example of such action might be

- refusal by the employer to allow certain rights, e.g., distributing recruiting literature, collecting subs, holding meetings in canteens and so on, to employees who belong to a union other than that recognised by the employer for that class of worker.

There is little difficulty in comprehending the first class of action; in these cases it should be straightforward for a tribunal to be convinced or otherwise by the employer that his reason for doing these various things was not connected with the right of the employee in respect of union membership or activities. A considerable difficulty, however, arises with the second class of action, which harks back to certain cases brought under the Industrial Relations Act in connection with employer-action short of dismissal. A key case of this nature was *The Post Office v Crouch*. This case was brought by an employee of the Post Office who complained that his Section 5(1) rights had been infringed. The background to the case was an inter-union battle between the Union of Post Office Workers (UPW) and the Telecommunications Staff Association (TSA). The UPW had sole negotiating rights for telephonists given by the Post Office. The latter refused, therefore, to give the TSA, of which Crouch was a member, the same range of facilities it gave the UPW. These facilities were precisely those referred to in (b) above, handing out leaflets, holding meetings and so on, and, claimed Crouch, amounted to discrimination against him and thus constituted an infringement of his right to take part in trade union activities. The case went all the way to the House of Lords; their Lordships held that while the UPW was entitled to more facilities than the TSA by reason of its sole negotiating status, the TSA was entitled to pursue activities which did not require the 'employer's assistance'. Thus they held that to discriminate against a union in respect of these 'non-assisted' facilities – which included recruiting members, collecting subs, handing out pamphlets and so on – was in breach of Section 5(1); it also constituted breach of Mr. Crouch's rights under that same section.

We return, then, to the Employment Protection Act. Dropping the term 'discrimination' is clearly an attempt to outflank the Post Office decision and allow an employee to show favour to a recognised union and hence, to show favour to employees who are members of that union even if by doing so they are (or would have said to have been) discriminating against employees who were members of a non-recognised, but equally independent, trade union.

Appropriate time. We saw earlier that an employee had a legal remedy against an employer who prevented him from taking part in the activities of an independent trade union at any appropriate time. The definition of appropriate time is the same as that used in the Industrial Relations Act. There are three broad classifications of appropriate time:

- time outside his working hours (working hours defined by reference to his contract of employment)
- time within working hours when by agreement it is permissible to take part in union activities
- time within working hours when by consent of the employer it is permissible to take part in union activities.

It should be noted that court decisions have established that time *outside* working hours, which requires neither the employer's consent nor agreement, includes time such as lunch hours when an employee may take part in trade union activities e.g., to attend a meeting or hand out recruiting leaflets in the yard; provided only that the activities in question do not put the employer to undue inconvenience, e.g., a mass meeting which prevents essential supplies being delivered during the lunch break.

It should also be noted that this right to take part in trade union activities differs from the right to take time off for trade union activities (*see Chapter 1*) in that the latter extends only to members of independent trade unions recognised by the employer for collective bargaining purposes. Taking part in trade union activities is not restricted to members of recognised unions for crucially such activity may include recruiting business. What may be galling for employers is seeing such business going on and being powerless to prevent it despite the conviction that in certain circumstances they are seeing a potential inter-union dispute brewing up before their eyes. This thought brings us to a limitation of the right to take part in trade union activities, which is where a closed shop exists.

Closed shop. Where there is a closed shop, trade union activity is restricted to members of unions specified in the closed shop agreement. This restriction applies only to activities on the employer's premises, which means that, say, recruiting can go on only outside the employee's working hours *and* outside the premises. This provision cuts out activities of a rival to the closed shop in the lunch break but does not of course prevent activities outside the gate once the hooter has blown. At the time of writing the definition of a union membership agreement, i.e., a closed shop agreement, is under discussion in Parliament in the context of the debate on the Trade Union and Labour Relations (Amendment) Bill.

Religious belief objection. It should be noted by employers that they cannot expressly take action of the type we have been discussing in this section, i.e. action short of dismissal, to compel a man to join a union if the man in question 'genuinely objects on grounds of religious belief'.

Union pressure to force an employer to take action short of dismissal against an employee. We saw earlier that the Trade Union and Labour

Relations Act deals with dismissal for trade union activities; that Act, however, gives protection to the individual against dismissal for objections to union membership on the ground of religious belief or any other 'reasonable' ground. At the time of writing the Trade Union and Labour Relations (Amendment) Bill is attempting to confine this protection to religious grounds only. It is obvious that there will be cases where the employer may be the innocent third party in a dispute between an individual and a trade union, and he may be forced by union pressure unwillingly to dismiss an employee. The effect of the Trade Union and Labour Relations Act, and the Amendment Bill when it becomes law, will be to withdraw the protection of 'unfair dismissal' from an employee who finds himself in this situation. What the Employment Protection Act does is to remove any grounds for a claim for compensation for action short of dismissal from such an individual.

It is arguable, of course, that an employer who gives in to such pressure is the victim of his own weakness and therefore has no real complaint if he is taken to a tribunal. Whilst this argument may have some force in certain situations, there have been a number of cases where employers, big and small, have been forced into making a choice between crippling industrial action or submitting to unions' demands to dismiss or discriminate against a man. A large firm may well be able to pay compensation in such cases though there is no reason in equity why it should. For small firms with limited resources however, compensation could be another matter altogether.

Compensation for breach of rights. An employee who feels discriminatory action has been taken against him by his employer on grounds of trade union membership or activities may complain to an industrial tribunal; this complaint must be lodged within three months of the action complained of (or within three months of the last such action if there has been a series of discriminatory acts). At the tribunal, it will be for the employee to show that his rights *have* been infringed; thereafter the burden of proof shifts to the employer who has to show that the purpose for which the action was taken was not to infringe the employee's rights. How the employer can do that where the action has been taken because the employee would not join a recognised union on the grounds of religious belief is left unexplained by the Act.

If a complaint arises on the grounds that a card-carrying Communist has been transferred from a design section producing guided missiles to another less secret section the obvious defence for the employer would be the interests of national security. The only conclusive evidence, however, a tribunal can legally accept in such a situation will be a certificate signed on behalf of the Minister of the Crown. No doubt it would be wise to obtain it before taking the action.

If the complaint is upheld, and the employee is able to prove that the action in question was taken against him, and the employer is unable to convince the tribunal that the action was taken for a reason other than the man's trade union membership or activities, a tribunal will award such an amount as it considers just and equitable in the circumstances and having regard to the infringement of the man's rights. It must also take into account such loss sustained by the employee as a result of the employer's action. This means that it is the intention of the Act not merely to compensate for loss, which may be negligible in money terms, but also for 'the general affront to an employee of having his fundamental rights of free association impeded', in the words of a Government spokesman who went on to say that 'this is comparable to the common law principle that compensation is payable for defamation'. No maximum is specified for the sum of such compensation under either classification. Employers will have to await cases to see what typical awards will be made under this heading of a trade unionist's rights, and therefore to assess what risks they may be running in this new industrial minefield.

CHANGES IN THE HEALTH AND SAFETY AT WORK, ETC. ACT 1974

The Health and Safety at Work, etc. Act 1974 represented a radical change in the legal structure relating to health and safety of people at the workplace. Essentially, it is enabling legislation, designed, eventually, to encompass all laws and regulations on health and safety matters, and it sets out general duties applicable to employers and employees. It gives wide powers to the newly formed Health and Safety Commission and Executive and to Health and Safety Inspectors. The Employment Protection Act makes three comparatively minor amendments to the Health and Safety Act, the most significant of which relates to employee safety representatives.

The other two changes deal respectively with the position of the agricultural industry, and powers of inspectors to disclose certain information resulting from enquiries and investigations at the workplace.

Health and safety employee representatives. Section 2 of the Health and Safety Act, which deals with the general duties imposed upon employers in relation to their employees, contained two specific provisions giving the Secretary of State power to issue regulations providing:

- that in certain specified cases, trade unions recognised by the employer should *appoint* safety representatives from among employees
- that in certain specified cases, employees should *elect* safety representatives from amongst themselves.

The safety representatives so appointed or elected would represent employees in consultations with the employer on health and safety matters (consultation through safety committees composed of safety representatives will also be required by regulations to be issued). In short, the intention was that safety representatives would have to exist in all companies – appointed by the trade unions in a unionised company or elected by employees in non-unionised companies.

The Employment Protection Act *deletes* that clause (2(5)) allowing the Secretary of State to require the election of employee health and safety representatives in the non-union or partly unionised company. Regulations, when they appear will thus only apply to the appointment of such representatives by trade unions where they are recognised by the employer. In practice of course, there is nothing to stop an employer from setting up safety committees with employee representation in the non-unionised plant or company and, indeed, there will be situations where it will be sensible and practical to do so. Difficulties are likely to arise however, where the employer recognises a union in respect of only some of his staff and this might only be a small proportion of the total. In such companies he may be left with the uneasy choice of either ignoring the bulk of his staff in respect of safety matters or, alternatively, setting up two committees – which might have conflicting views on some issues.

This amendment is in line with general TUC pressure for all channels of consultation, representation and participation to be 'single channel' namely the channel of trade unions. There are those who would feel, and the view has been sharply expressed, that in the particular context of health and safety it is unfortunate that political dogma should override the general concern for an improvement in health and safety standards at work.

The agricultural industry. The Health and Safety Act made one specific exception to its general purpose of bringing all health and safety matters across industry under the general control of the Health and Safety Commission. In many significant respects, responsibility for health and safety matters in the agricultural industry rested with the Ministry of Agriculture, Fisheries and Food. This came about because of the special nature and problems of that industry, and particularly, the number of small units within it. The Government subsequently changed its view and as a result, the Employment Protection Act amends the Health and Safety Act so as to bring the agricultural industry within the ambit of the Health and Safety Commission. It does however, allow the Minister of Agriculture to act independently on certain minor matters and provided that there has been prior consultation with the Commission. There is also provision for the Minister of Agriculture and the Commission to act jointly in certain cases.

Disclosure of information. Section 28, subsections 7 and 8 of the Health and Safety Act effectively preclude Health and Safety Inspectors from disclosing information obtained as a result of official investigations and enquiries relating to health and safety matters, e.g., an accident at a factory – except

- for purposes of his functions; or
- for purposes of legal proceedings; or
- with the consent of the person responsible for the premises in question; or
- for purposes of keeping employees at the workplace adequately informed about matters which affect their health, safety and welfare (providing that the information is factual

and relates to the employer's premises. It may also be given to the employees' representatives).

It was subsequently pointed out to the Government that the Act as drafted precluded inspectors from carrying out what had been an established practice – namely the ability to provide factual information to an employee and/or his solicitor where that employee was likely to be party to legal proceedings. The Employment Protection Act has therefore added another subsection to section 28 enabling this practice to continue.

BREACHES OF CONTRACT OF EMPLOYMENT

To date, claims relating to breaches of the employment contract have been dealt with by the County Courts or by the High Court. The Employment Protection Act, in line with recommendations made by the Donovan Royal Commission as long ago as 1968, gives the Lord Chancellor power to confer jurisdiction on industrial tribunals to hear certain claims relating to breaches of the employment contract. Pursuing claims through the ordinary courts is, for the average employee, an expensive and time-consuming business. Industrial tribunal proceedings are considerably less onerous in both respects and it therefore seems not unreasonable to predict an increase in the number of breaches of contract claims once the tribunals are granted the appropriate jurisdiction. However, since the tribunals will take some time to digest the enormous increase in workload occasioned by the Health and Safety Act 1974, the Sex Discrimination Act 1975, the Trade Union and Labour Relations Act 1974, and of course, the other sections of the Employment Protection Act, it may be some considerable time before an order conferring jurisdiction on the tribunals in respect of breaches of contract is made.

As we have already mentioned, the type of breaches of contract cases which may be heard by the tribunals will be limited. Only claims which:

(a) arise, or are outstanding on termination of the employment contract; or

(b) relate to proceedings already or simultaneously brought before the tribunal under some other section of the Act, or any other Act over which the tribunals have jurisdiction (e.g., an unfair dismissal case);

may be heard by the industrial tribunals. Of course, a claim for breach of the employment contract which does not fulfil the above conditions may continue to be dealt with by the ordinary courts.

A claim which meets the appropriate conditions may only be a claim for damages as a result of a breach of contract, or a sum due under the contract, or a sum due under legal provisions relating to any contract of employment (e.g., maternity pay). In all such claims the industrial tribunals will have power to order the employer to pay the claimant any amount which it finds due. There is one specific exclusion – and that is damages relating to personal injuries.

The employer would be well advised to use the period before this section of the Act comes into force to review his employee's contracts of employment:

- to ensure that the terms of the contracts are clearly laid down in writing, or known to the parties *(see below)*; and
- that changes in terms of the contract are carefully considered and carried out in the correct manner.

It is important, too, that the nature of the employment contract is clearly understood by administrative and personnel departments. The Contracts of Employment Act 1972, does not of itself define the contract of employment. It merely requires *inter alia* that certain specified terms, such as 'hours of work' be set out in a written statement. There may, however, be terms of the employment contract which may not be written down either in the required statement or in any other document – but which, nevertheless are implied terms of the contract. Such an implied term could for example be a 'custom and practice' arrangement relating to regular overtime payment.

The order relating to breaches of contract claims, when it is made, should be carefully examined, since the Lord Chancellor has power to vary or restrict the range of claims which may be brought to the tribunals and also to include provisions as to the manner and time in which such claims may be made.

Part II:
Collective Bargaining

Chapter 4: Recognition

Writing in the *Financial Times** on the Employment Protection Bill as it then was, the paper's Labour Editor, John Elliott, stated: 'In their anxiety to gain full statutory backing for the many advances contained in the Bill, the unions have in fact designed a new framework of law to replace the Conservative one which they demolished with such invective and glee last year. It will be a union-favoured framework. But on items such as trade union recognition, collective bargaining and disclosure of information and pay arbitration, the framework could well lead to quasi-legal conflicts with appeals to the formal courts of law and with Law Lords handing down judgements which could restrict union operations. In addition the Bill provides a ready-made basis on which a future Conservative or any other government could build restrictions to correct the shift in the industrial power balance which employers expect the Bill to cover.'

In the series of four chapters which follow we shall describe those sections of the Act which can conveniently be categorised as having some effect on the conduct and outcome of collective bargaining and also attempt – though it is a difficult and speculative task – to estimate what changes will come about in collective bargaining as a result of the passage of the Employment Protection Act on to the Statute Book. In his text on the Industrial Relations Act 1971† Peter Paterson said: 'It would take a soothsayer to predict precisely what impact the new legislation will have on British industry. Clearly, the imposition of a fairly elaborate legal system on an area which has traditionally tried to keep well clear of the law in solving its problems will make some difference.'

This comment, as events turned out, was most far-sighted. And it can certainly be applied to the Employment Protection Act which provides for recourse to the law, though not inevitably so, in areas which have always been, except for the transient effect of the Industrial Relations Act, the subjects of collective bargaining. This

*11 April 1975.

†*An Employer's Guide to the Industrial Relations Act*, Kogan Page, 1971.

is so in respect of the many individual rights extended to employees by the Act (which will be described in Chapters 1–3) but it is crucially so in the conduct and practice of collective bargaining.

One major difference between the Employment Protection Act and the Conservatives' Industrial Relations Act is that the trade unions have actively sought the changes and additions to the law which have now been brought about. Therefore employers must expect that the TUC and individual unions will encourage the members to take up their individual rights, and, crucially for the next chapter dealing with collective bargaining, use the sections of the Act which give trade unions more rights. All this is very different from the situation following the passage of the 1971 Act where not only were unions actively discouraged from using their rights under that Act (a number of unions, for example, were actually expelled from the TUC for registering) but militants looked with disfavour on individuals exercising rights such as the right not be to unfairly dismissed (the latter probably only in the very early days). Thus employers must expect the worst.

RECOGNITION OF TRADE UNIONS

As many employers know to their cost, the process of recognising a union or unions to represent groups of employees can often pose complex problems. Even where the employer is ready and willing to negotiate with a union, he may find that several unions are competing for recognition or that existing bargaining arrangements are being challenged by another union. One of the objects of the TUC's Bridlington Agreement is to try to discourage its affiliated unions from 'poaching' on each other's territories. The need for an independent body to consider such problems has, moreover, been recognised for some years. The Industrial Relations Act of 1971 attempted to do this with the establishment of the Commission on Industrial Relations, and the National Industrial Relations Court to which recognition issues could be referred by employer or union. The fact that comparatively little use was made of these provisions must be measured against the total opposition of the TUC, on principle, to the whole of the Industrial Relations Act.

The new Act, to an extent, replaces some of these provisions, but whilst the object may remain the same, the approach is very different. It would not be unfair to say that, generally, the Act's provisions on recognition are there for the use of those unions who care to use them, and we can envisage situations in which the employer would not be averse to their doing so. Nevertheless, there are ways in which the employer's view may be taken into account once an enquiry has been initiated, as will be apparent by the following description of the Act's provisions on recognition.

One or more independent trade unions (but no employer) may refer a recognition issue to the Advisory, Conciliation and Arbitration Service. It is worth noting that a 'recognition issue' does not only mean primary recognition for negotiating purposes, but can also mean an extension of recognition rights* for a union with whom the employer already negotiates. For example, a union which has negotiating rights for clerks, may ask for recognition to be extended to supervisors in the same company. It also seems possible for a union recognised by one employer to seek to extend that recognition to an associated company.

Of perhaps even greater significance is the possibility of unions using these procedures to try to extend the scope of collective bargaining to cover matters which the employer considers to be his prerogative. The attitude to be adopted by ACAS towards any such ambitions remains to be seen.

If ACAS does not turn down a claim for recognition on the grounds that it has already been the subject of previous reference, and that no significant change has since occurred, it will consult all parties whom it considers will be affected by the outcome of the reference. Such consultation will, of course, include the employees concerned, but ACAS is not obliged to ascertain their opinions by means of a ballot.†

Where a ballot is used, the Act requires that ACAS ensures that every employee concerned has a fair opportunity to vote, and that the ballot is secret. Before the ballot takes place, ACAS must notify

*The phrase might also mean trying to extend the matters presently covered by agreements to, say, financial planning or investment decisions; an amendment clarifies this so that recognition can only be given for matters regarded as potential subjects for trade disputes as defined in the Trade Union and Labour Relations Act.

†*Vide* Kessler and Palmer: *Personnel Management*, July 1975. 'ACAS, for example, is given an extremely free hand in deciding recognition claims. The legislation does not specify the enquiry method, nor the criteria ACAS should use to reach its judgements, yet those judgements in the last resort will have legal effect.'

the employer and unions concerned, specifying the categories of employees to be included, and the questions to be asked. At this stage, the employer, or the unions, may make representations to ACAS on any of these points. The Act does not specify the amount of support which would qualify for recognition whether or not a ballot is used, but one must assume that common sense will prevail, and that recognition will not be recommended against the wishes of a substantial number of employees.

Having made appropriate enquiries, and assuming the matter has not been settled by voluntary agreement between the parties (which ACAS is under a statutory duty to promote) ACAS will produce a written report, either recommending recognition or not, and stating reasons in either case. A positive recommendation must identify the employer and unions and descriptions of employees concerned, and the extent and level of recognition. It may also specify that certain conditions be complied with by the union before the recommendation takes effect. What such 'conditions' might be are not specified, though they could be crucial to the whole issue. Presumably they might include the condition that agreement is reached between competing unions in which case, hopefully, the power to delay the date on which a recognition award becomes effective until specified conditions are met could act as a spur to such agreement. Such 'conditions' might also, for example, include a requirement that the union provide an adequate number of qualified officials to service the agreement.

Once a recognition recommendation has become operative, it remains so unless superseded by agreement between the relevant parties, or by another recommendation, or if revoked on application to ACAS.

The employer has no power of appeal against a recommendation but he may, as may the union concerned, apply for it to be revoked. This applies only, however, if he can show that circumstances have changed sufficiently since the recommendation was made, to justify reconsideration. Effectively, then, neither employer nor union have any right of appeal in the event of disagreement with an ACAS recommendation. What happens, then, if either party refuses to comply?

There is no legal sanction which the employer can resort to in the event of non-compliance on the part of the union, even where non-compliance amounts to industrial action. The union, on the other hand, although not given the right to enforce recognition directly, is provided with a powerful sanction. Once a recognition recommendation becomes effective, which is 14 days after the employer has been notified of the recommendation, then the employer is given two months in which to comply. Should he not do so then the union can complain to ACAS that the employer is not complying with the recommendation. Should conciliation fail, the union may then refer the matter to the Central Arbitration Committee (CAC) including specific reference to terms and conditions forming part of the claim on which in its view, the employer is failing to negotiate in a manner consistent with effective recognition of the union. (Such a complaint will, however, be turned down if a similar complaint had been made, and an award made, in the previous twelve months. The CAC will then have the power to arbitrate unilaterally on the terms and conditions specified in the union's claim, and its award will become part of the relevant employees' contracts of employment replacing any terms to the contrary.

It is, of course, only where a recognition recommendation has been imposed by ACAS that the union may make such a complaint to the CAC. A claim on terms and conditions often follows closely on the heels of a recognition agreement, if it has not already preceded it, but where the ACAS procedure has been invoked, this will almost certainly be the case. Refusal to consider the claim, or slowness in replying on the employer's part (especially in view of the two months' time limit for the union to complain), could well constitute grounds for a CAC award.

It is not unreasonable to assume that unions will make greater use of the facilities outlined above than was made of similar provisions in the old Industrial Relations Act, to which they were so bitterly opposed. In reviewing existing or anticipated collective bargaining arrangements, employers will be well advised to be aware of these new rights available to trade unions and take them into account in formulating policies and strategy. Codes of Practice, to be issued by ACAS, may clarify its policy towards recognition issues generally, and to the extent that they do, will make a useful contribution towards the framing of company policy on recognition arrangements.

Employers' organisations have consistently complained that employers do not have access to the new procedures – a disability which will be particularly important in inter-union disputes. The Government's reply to them has been that if employers were allowed to initiate complaints to ACAS, such a procedure would only be sensible if the CAC decision had some sort of sanction available (as, of course, it has with a union-initiated complaint, *viz.* an order to incorporate claims as implied terms of contract). In the case of employer-initiated complaints, a sanction might be introduced making it unlawful for the losing unions to strike in protest. Not unnaturally, this Government is not prepared to contemplate such action and thus they are not prepared to contemplate employers having access to the procedures outlined above. In the event, employers will have to rely, in inter-union membership rows, upon the Bridlington Agreement and *(see Chapter 9* – ACAS) the power of ACAS to initiate an enquiry and produce a report – the findings of which would have no legal authority. It is little wonder that Michael Foot, the present Secretary of State for Employment said, in welcoming the new Act, that it would help to give workers 'a chance to share control of industry'.

RECOGNITION ISSUES

(A) Procedure for handling recognition issue reference to ACAS

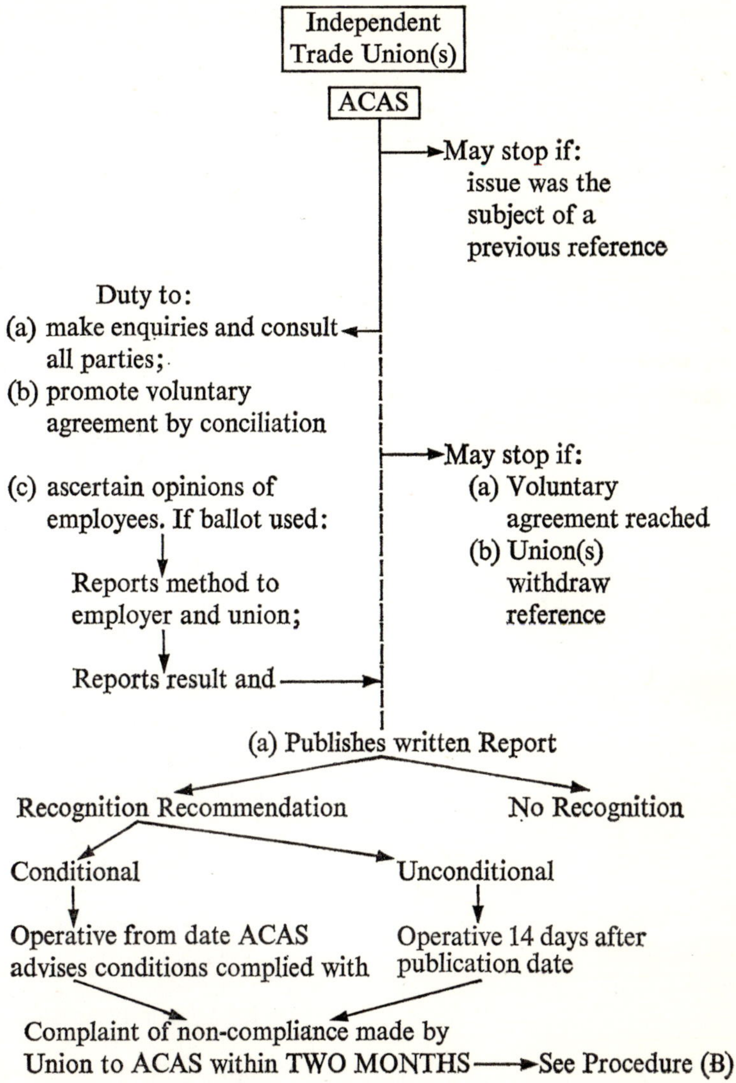

RECOGNITION ISSUES

(B) Procedure for handling Union complaint on non-compliance by Employer with Recognition Recommendation.

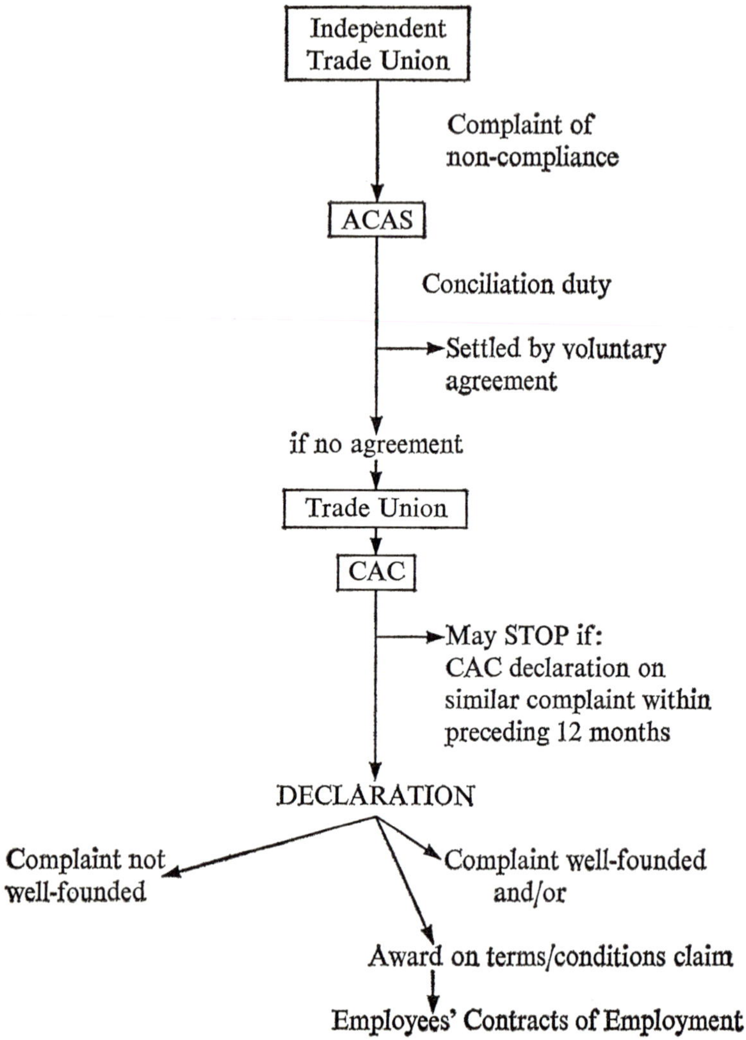

Chapter 5:
Disclosure of Information

DISCLOSURE OF INFORMATION

It will be recalled that the section of the Industrial Relations Act which had as its intention an obligation on companies to disclose to trade union representatives information necessary for the purposes of collective bargaining was never activated. It aroused a great deal of comment in 1971 – much of it hostile comment from employers – but it was also noted that it was not held to be an unfair industrial practice for an employer to refuse to disclose. Instead, the Industrial Relations Act prescribed that trade unions which suspected that they were not being given the information they required could seek a remedy through the National Industrial Relations Court. That body could make an order determining the rights of the (registered) unions and direct the employer to take whatever action he ought to be required to take. A further sanction of the Court would have been to authorise the union to take its claim (i.e., the claim for improved wages or conditions for which the information about the company had been required) to the Industrial Arbitration Board, which could then decide on the merits of the claims and order the employer to pay up – the award being back-dated to the date on which he was in breach of his duty to disclose the information.

If we substitute the Central Arbitration Committee for the National Industrial Relations Court and Industrial Arbitration Board, then Section 56 of the Industrial Relations Act (along with assorted bits) issues from the corpse of the Act that never quite did give it birth, to emerge finally into the world by the agency of the Employment Protection Act. So disclosure should go down in the form book as by Carr out of Foot.

In sequence the procedure laid down by the Employment Protection Act is as follows:

(a) a general duty is placed upon all employers (but not employers' organisations, which exclusion rules out the possibility of legal demands for disclosure of information in national agreements) to disclose on request to trade union representatives *all* such information relating to his undertaking or that of any associated employer*, which information is both:

(i) information *without which* the trade union would be 'to a material extent' impeded in carrying out collective bargaining *and*

(ii) information of a type which should be disclosed in line with good practice (on which good practice the relevant section of the Code of Practice may be prayed in aid – but not exclusively so. This point emphasises the importance of the duty given to ACAS to produce a new Code of Practice on disclosure.)

(b) the right to such information resides with independent trade unions which are recognised by the employer for the class of labour to which the claim refers. Additionally such right is possessed by a union which has been given an ACAS recommendation for recognition in accordance with the procedure described earlier in Chapter 4.

(c) As in the Industrial Relations Act there are certain get-outs for the employer. He need not disclose:

(i) information, which if released, might damage national security (on which the evidence of a Minister would be regarded as conclusive)

(ii) information which he could not disclose without contravention of an enactment

(iii) information which has been communicated to him in confidence

(iv) information about an individual (unless the individual's consent has been obtained)

*This information relating to the employing company but in the hands of a holding company would have to be disclosed, but not necessarily information about another associated *company*.

(v) information of such a nature that its disclosure would be seriously prejudicial to the interests of the undertaking – but not if that prejudice to his interests was solely in the context of collective bargaining. (For example it might be prejudicial to the employers' interests to disclose his breakdown of costs because a competitor could pick this up; he will not be absolved for releasing however, say the distribution of (unnamed) individuals by salary in a grading system simply because that disclosure might prejudice his interest to the extent of making a claim for rationalisation of that distribution by a trade union almost inevitable)

(vi) information obtained by him for the purpose of taking legal action

(vii) the employer need not permit inspection or copy of original documents

(viii) the employer also need not compile information where he considers the effort in making such a compilation would be out of all proportion to the value of the information (to the trade union) in the conduct of collective bargaining.

In the event of an employer having to defend himself against a complaint of non-disclosure, the onus of proof would be on him to sustain these exclusions. No doubt there will be some entertaining evidence – entertaining that is for spectators – produced in regard to no. viii of the possible exclusions.

(d) If a trade union considers that an employer has not disclosed information it may complain to the CAC which, if it thinks that the matter may be settled by conciliation, will push it on to ACAS to see if it can do that. Presumably if the employer accompanies his refusal with remarks like 'over my dead body', etc., CAC may skip this step.

(e) ACAS will then inform CAC in the event that it cannot settle the matter by conciliation.

(f) CAC will then hear and determine the complaint and give

its judgement; any person whom CAC considers has a proper interest in the complaint will be entitled to be heard (say an individual who would be named if certain information was given and who had not given his consent) as, of course, would the employer and the trade union making the complaint. If CAC finds the complaint justified then it will:

(i) detail the information not disclosed

(ii) confirm the date on which the employer failed to disclose

(iii) state a period in which the employer ought to disclose the information (such period being not sooner than a week following the judgement).

(g) If the period specified by CAC for disclosure expires, then the trade union in question may make a further complaint to CAC. That body will then conduct another hearing and adjudge accordingly. When it presents this further complaint, the trade union may (and almost certainly will) present to CAC a substantive claim on wages or conditions or both, which it reckons should apply to the group of workers (this in practice would normally, though not always, be the claim which started off the whole proceedings) in question. One could imagine no doubt that the original claim had been submitted quite a few months before the somewhat tortuous procedure outlined above had been completed; inflation being what it is, it would no doubt be in order for the trade union to revise it upwards.

(h) CAC may then, as a remedy, award the claim or such part of it as it considers appropriate and that award will have effect as part of the contract of employment of each employee covered by the claim.

(i) Finally, if a recalcitrant employer (who no doubt has still not given the information and does not intend to) does not adjust his wages and conditions in accordance with the CAC award, then the individual employees will be obliged to take their case to an industrial tribunal for a further judgement; the Act may empower tribunals to adjudge upon claims for damages for breach of a contract of employment *(see Chapter 3)*.

This involved and lengthy procedure arises from the simple fact that it is very difficult to think up alternative sanctions to apply to an employer who does not disclose information. No doubt it is hoped that in the course of the process he will come to the conclusion that enough is enough, and disgorge anyway.

Finally it might be said that the provisions on disclosure of information in the Act strike us as not only clumsy but really rather old-fashioned. A new spirit is moving among employers in industry which is beginning to regard information on companies' activities not as something to be 'disclosed' – a term which implies extreme reluctance – but something to be imparted with skill and conviction. It is to be hoped that these provisions in the Employment Protection Act will not unduly hinder that new spirit which, together with moves towards workers' participation, represents a most important trend in British industry – a trend which could lead to constructive involvement of employees in the enterprises in which they work.

GUIDELINES FOR EMPLOYERS ON DISCLOSURE OF INFORMATION

Recognise that 'disclosure' of information is a valuable management tool and not simply a union weapon. Institute a positive information policy and apply it (see Appendix for a CBI guide to information disclosure).

Be ready for union claims in this area by preparing sensible views on:

- 'impeded to a material extent'; in an extreme case the demolishing of a union contention that it was materially impeded in pursuing a simple wage claim because an employer would not divulge his forward investment plans
- 'information seriously prejudicial to the interests of the undertaking'
- compiling of information where the effort involved would be be 'out of all proportion' to the value of the information to the trade union in collective bargaining.

If a union enlists outside help (e.g., Ruskin College) to mount a case in this area, remember it may very well just be on a 'fishing' expedition and the real target is not you but another employer. Consult an appropriate employers' organisation in such a case.

Chapter 6:

Disputes and Strikes

This chapter continues the examination of how the Employment Protection Act will affect collective bargaining. We look at the objectives and functions of the Advisory, Conciliation and Arbitration Service – whose tentacles reach into so many aspects of the new legislation – and the changes to the Social Security and Supplementary Benefits Acts which will make it easier for workers indirectly involved in industrial disputes to obtain unemployment pay or supplementary benefit.

ADVISORY, CONCILIATION AND ARBITRATION SERVICE (ACAS)

Assiduous followers of the legislative scene and industrial relations will recall that in 1968 the Royal Commission on Trade Unions and Employers' Organisations, under the chairmanship of Lord Donovan, recommended that a permanent Commission on Industrial Relations should be set up. The Labour administration of the day accepted the recommendation, and the Commission was set up in 1969 by Royal Warrant. In 1971 the Industrial Relations Act confirmed the Commission on Industrial Relations on a statutory basis and in 1974 when the political wheel had turned again the CIR was abolished by the Trade Union and Labour Relations Act 1974. However, no time was lost in setting up the Conciliation and Arbitration Service also in 1974. This body – referred to in the Employment Protection Act as 'the Service' is, like the CIR before it, accorded statutory recognition by legislation. The Advisory, Conciliation and Arbitration Service as it is now known is directed by a council of a tripartite nature* consisting of a full-time chairman and three members appointed in consultation with 'such organisations representing employers as the Secretary of State considers appropriate' (presently the CBI), three members appointed in

*See Appendix 1 for further details on ACAS.

consultation with 'such organisations representing workers as the Secretary of State considers appropriate' (presently the TUC) and three independent members appointed by decision of the Secretary of State. There is also provision for appointing up to three full-time or part-time deputy chairmen to the Council – who may not necessarily be additions but who may be appointed from the ranks of the nine.

This council then is required to direct ACAS. We shall now examine what it is that ACAS is required to do by the Act and what likely effect the discharging of these functions may have upon that hardy plant, collective bargaining. In general terms ACAS is required to:

– promote the improvement of industrial relations

– encourage the extension of collective bargaining

– develop collective bargaining

– reform, where necessary, collective bargaining

These duties imply very clearly that ACAS, whatever else it is there for, has to devote itself to extending and developing collective bargaining; this may have been tacit public policy before; in the reports of the CIR operating under a Conservative Government it is not difficult to discern a bias in favour of extending recognition to trade unions and enlarging the scope and coverage of collective bargaining. But the appearance of these duties in the Act means that this tacit policy has been turned into law. Evidently those employers who, on whatever grounds, resist recognising trade unions* can expect no sympathy from ACAS. Such a situation may very well bring stresses upon the employer representatives on the council of ACAS; particularly when ACAS comes to revise those parts of the Code of Practice relating to disclosure of information by employers in collective bargaining. The inherent contradiction in their role is that they are part of a body with such union-biased duties given it by Parliament.

Let us look more closely at the functions recognised in ACAS's title.

Conciliation. ACAS has inherited the administration and personnel of the Department of Employment conciliation service which dates from

*See discussion in Chapter 4 – Recognition.

the Conciliation Act of 1896; their experienced staff retain their status as civil servants but their role is intended to be seen as independent of Government policy. Such independence, in the intention of the framers of the Act, is more clearly seen because the conciliators will be directed, not as heretofore, by a Government department but by the Council of ACAS. The staff, however, like the rest of the ACAS staff, are civil servants which may, to unkind eyes, render their claim to be independent a little suspect. It is likely that the unions' eyes will be unkindest of all in this regard. There are four duties specified for the conciliators in the Act.

Dispute settlement. The ACAS conciliators are obliged to respond to a request by one or more parties to a trade dispute (or parties to one that looks likely to happen) to offer their services to try to bring about a settlement. This means that if for example a union represented in a plant sees that it is heading for a clash on wages with the employer, then it can ask ACAS to help keep trouble away; alternatively if an employer is in the middle of a dispute with his unions, on say his proposals to get rid of piecework and substitute measured day work, then the employer can call in ACAS to try and sort out the dispute. Interestingly, if neither party asks for help ACAS has the option of offering its services anyway if it so desires.

This conciliation function is well enough established; it is not expected that its inclusion under the administrative wing of ACAS will make much day to day change in these activities. Conciliation as well as arbitration is often described as a third party trying to buy peace at the lowest possible price, and while this ideal may suffer some shocks during periods of incomes policy it is one which because of sheer need, will have to go on. Little change then in this area. As before, conciliation officers will try to see that agreed procedures are followed – despite the fact that they may be called in only because these procedures have been flouted.

Furthermore, ACAS will be obliged to take over conciliation in cases of complaints of unfair dismissal. In these cases a copy of the complaint of unfair dismissal, together with the employer's reply must be sent to the conciliation officer and the parties must be notified that the services of the conciliation officer are available to them. Normally the tribunal hearing is not held until the conciliation officer has tried to arrange a settlement – which alas for the employer is often a suggestion by the conciliator that he settle for an agreed sum.

It should be noted that ACAS is limited to conciliating in 'trade disputes' which, although widely defined under the Trade Union and Labour Relations Act 1974, do not include disputes of a political nature, such as an embargo on goods destined to Chile or Spain; the definition also does not include disputes concerning workers abroad unless workers in this country will be affected by the outcome.

The conciliation officers will also deal with complaints of breaches of the new individual rights *(see Chapters 1, 2 and 3)* conferred by the Employment Protection Act and also complaints relating to employment under the Sex Discrimination Act.

Arbitration. In cases of trade disputes where conciliation has been tried (or not tried because it seemed unpropitious) and has been unavailing, and where one of the parties to the dispute requests arbitration, ACAS may, but is not obliged to, refer the matter to arbitration. In any event there will be no use of arbitration unless *all* parties (i.e. normally the union *and* the employer) agree to arbitration.

The choices of arbitration mechanism are:

(a) one person appointed by ACAS (not an officer of the service – in fact likely to be one of the panel of academics who hold themselves available for such tasks)

(b) more than one arbitrator – generally three – appointed by ACAS; in this case ACAS will nominate one of them as chairman

(c) The Central Arbitration Committee.

Because the consent of all parties is required for arbitration, it can be deduced that consent is also required as to the form of arbitration chosen, as well as the persons appointed by ACAS. Consent by all parties is also required to make the award binding, as well as to whether or not it is published. It should be noted that arbitration awards in labour matters have never been viewed as legally enforceable and are specifically excluded from the Arbitration Acts. Thus, an agreement to arbitrate and to accept the award as binding would not be legally enforceable. It might also, however, be viewed as a collective agreement if made between a union and an employer or employers and would then only be legally enforceable if it were in writing and specifically stated that the parties intended it to be legally enforceable as provided in the Trade Union and Labour Relations Act 1974. This does not apply where the arbitrator or CAC is acting in pursuance of some statutory authority which makes the award binding upon the employer, e.g., an award under the recognition procedures *(see Chapter 4)*.

It has always been a source of puzzlement to foreign, particularly US, observers of the UK industrial relations scene that arbitration is not more widely used here. In part this is because the distinction so familiar to participants in other collective bargaining systems, between disputes of right and disputes of interest, is not generally recognised in practice in Britain. In brief, a dispute of right is a dispute about the interpretation of an agreement whereas a dispute of interest is one which takes place before the agreement is concluded. The distinction is easier to observe when labour agreements are legally binding which they are generally not in Britain, but the fact remains that arbitrators in Britain are asked to deal with both types of disputes or even a dispute which is a mixture of both. When British arbitrators are dealing with matters of interest they generally give a decision but no reasons and employers, at least in the last fifteen years or so, have shown great reluctance to have amateurs decide matters of vital economic interest to them, such as wage rates and conditions. When that has happened it has generally done so in an atmosphere of last-ditching and it is difficult to see anything in the arbitration processes offered by the ACAS which will change that. Employers should remember that while conciliation officers can be called in at the request of the unions, arbitration can only take place with the consent of both the union and themselves.

Finally, we may note that clearly arbitration, and to a lesser degree, conciliation, in the context of counter inflation policies will bring problems for ACAS. It is ironic to recall that the then Ministry of Labour in its written advice* to Donovan examined the possibility of removing the Conciliation Service from the Ministry and transferring it to an independent agency. It was stated:

> 'One argument advanced is that if a dispute should arise over a claim which appears to conflict with the Government's incomes policy, and one or both of the parties to the dispute should seek conciliation from the Ministry of Labour, it is

*HMSO 1965.

difficult for the Minister and his officers to intervene if this could lead to a settlement in conflict with Government policy.' The Ministry went on to say in rejecting such an argument for setting up an independent service that:

> 'An argument against such a change would be that the Government cannot delegate the ultimate responsibility for industrial peace and that it is therefore sound for the Government to retain direct control over the Conciliation Service.'

Alas, Governments have ultimate responsibility not only for industrial peace but also for keeping wage-led inflation within bounds. This conflict of interest has really only been pushed aside by the establishment of ACAS and may emerge again into public view before long.

Advice and enquiries. When the CIR was established as a statutory body by the Industrial Relations Act it retained its previous investigative, non-executive responsibilities, but its inquiries into the state of industrial relations had to be initiated by the Secretary of State or by the NIRC. ACAS is considerably more independent in this field; if it thinks fit – by which we could assume its Council thinks fit – the Service may offer advice on *any* matters concerned with industrial relations or employment policies or enquire into *any* question relating to industrial relations generally or industrial relations in any particular undertaking.

From the days when the CIR was established on a Royal Warrant basis it seemed obvious that the Secretary of State and his department were reluctant to provide it with references which might prove of a controversial character. No major car assembly firm for example was referred to the CIR though observers might be forgiven if they thought that most of the unfortunate features of British industrial relations were well represented in that industry. No doubt this coyness was due to the fact that both managements and unions in embattled areas tend to be particularly reluctant to let in outsiders to investigate their problems. One can understand why. Not only might problems be highlighted which both sides would prefer to see kept out of sight, not simply because of fear or idleness, or in pursuit of that much-admired industrial relations policy of letting sleeping dogs lie, but because the inexperience and possible prejudice of the enquirers might do more harm than good. These objections might equally apply to the advice and inquiry obligations of ACAS and it is an interesting speculation as to whether that body will be more successful than its predecessors in getting to grips with those obligations.

Employers will have to prepare themselves for two specific situations where ACAS decides to offer advice or institute an inquiry.

First, ACAS may, if it thinks fit, on request or otherwise, provide to employers, employers' organisations, workers and trade unions such *advice* as it thinks appropriate on any matter concerned with industrial relations or employment policies, including, but not exclusively, the following:

- the organisation of workers or employees for the purpose of collective bargaining
- the recognition of trade unions by employers
- machinery for the negotiation of terms and conditions of employment and for joint consultation
- procedures for avoiding and settling disputes and workers' grievances
- questions relating to communication between employers and workers
- facilities for officials of trade unions
- procedures relating to the termination of employment
- disciplinary matters
- manpower planning, labour turnover and absenteeism
- recruitment, promotion and vocational training of workers
- payment systems, including job evaluation and earned pay.

ACAS may also publish general advice on any matter concerned with industrial relations or employment policies; presumably specific advice directed to an employer or a union would be regarded as not general enough to warrant publication.

Secondly ACAS may, if it thinks fit, enquire into any question relating to:

- industrial relations generally
- industrial relations in any particular industry
- industrial relations in any particular undertaking
- industrial relations in any part of a particular undertaking.

Thus ACAS might, of its own volition, enquire into, for example:

- the adequacy of written agreements at plant level
- industrial relations in the motor industry
- industrial relations in Ford's
- industrial relations in a particular plant of Ford's.

The enquiry will produce a report which will constitute both findings and advice connected with these findings. Unlike the general advice previously referred to (though not specific advice) such findings and related advice may only be published by ACAS if it appears to that body to be desirable for the improvement of industrial relations, either generally or for the specific matter which was the subject of the enquiry. ACAS may only publish if it sends a draft to the parties concerned and taking the views into account (which views may include the view that the findings and advice should *not* be published).

ACAS has no legal powers to compel production of documents or subpoena witnesses, and it will be vitally dependent upon employer co-operation in particular. Most employers in these circumstances will be wary of taking the line of non co-operation because of fear of labour unrest, bad publicity and the like. A better plan might be to draw up plans for taking the fight to ACAS on terms of reference, and lobbying, if necessary with ACAS council members, to obtain terms of reference which will include things the unions would not like to see exposed to the (possibly) public gaze.

Another situation employers may find themselves in, is trying to decide whether they should lobby for the receipt of advice or enquiries from ACAS. ACAS, of course, has complete discretion as to whether to give advice or hold an enquiry. Situations in which they might wish to get advice or have an enquiry mounted are not too difficult to envisage, e.g., persistent flouting of procedure by unions and their representatives, insistence upon the maintenance of restrictive practices, inter-union difficulties, and so on. The latter will really cause difficulties for ACAS, for as we have seen in Chapter 4, the Act does not allow access by employers to ACAS on recognition questions directly; in an inter-union dispute it is highly likely that none of the unions will wish to use the procedure available to them in the Act. In these circumstances seeking an enquiry by ACAS might be sensible strategy for an employer – even though the subsequent report would not be legally binding. An employer could not, of course, get ACAS to hold an enquiry if it did not think fit; and ACAS may shy away from inter-union disputes, preferring to rely on the TUC's procedures.

It would appear that the only minor chink in the quite extraordinarily wide scope given to ACAS in the matter is, as we have seen, that it cannot publish the findings of an enquiry unless it takes into account the views of all parties concerned. 'Taking into account' however is not the same as requiring consent, and it may be that this phrase will make employers more cautious than ever in extending whole-hearted co-operation to the ACAS enquiries.

ACAS, we deduce, is an odd body. It has wide powers under legislation and an even greater discretion. It is made up of civil servants from the Department of Employment conciliation, arbitration and advisory services, and it has enquiry branches staffed extensively, but not exclusively, by ex-CIR staff. It is to be directed by a largely part-time Council which has the authority to determine its own procedure and is, into the bargain, a quite delicately balanced group of competing interests. It is charged quite specifically with tasks which many employers will find distasteful and provocative. Little good will be done for industrial relations if ACAS does not operate its function with tact and discretion. We hope it does and we must add that the experience and character of its present Chairman gives much encouragement that it will.

Codes of Practice. It may be noted in the schedule to the Act which deals with transitional provisions, that the Code of Practice in the Trade Union and Labour Relations Act (which was originally inspired by the Industrial Relations Act) is to remain in effect and accordingly will continue to be taken into account, where relevant, in tribunal cases, and in proceedings before the CAC. However, ACAS may issue what are described in the Act as new Codes of Practice. As we have seen it is specifically charged to issue new Codes on disclosure of information by employers to trade union representatives for the purpose of collective bargaining, and on giving time off for trade union officials.

However it is obvious that the 1971 Code of Practice is not intended to survive only with the amendment of the sketchy guidance on disclosure of information and the addition of guidance for time off from work for trade union representatives. In passing, it might be observed that para 29 of that Code states:

> 'As a first practical step, management should make available, in the most convenient form, to recognised trade unions, the substance of information which is supplied to shareholders or published in annual reports.'

This represents a statement of principle which can hardly be regarded as wildly progressive in the mid 1970's.

Clearly, then, ACAS will be getting on with re-writing the Code in batches; employers should remember that ACAS is obliged to prepare a draft of any proposed changes to the Code and consider such representations as may be made to it about the draft. Thereafter the possibly modified draft will go to the Secretary of State who, if he approves it, will send it to both Houses. Thereafter Parliament has 40 days to object; if it does not, the Secretary of State will put the revised portion of the Code into effect by order. To hazard a few guesses, priority areas for revision are likely to be: recognition, grievances and dispute procedures, disciplinary procedures, responsibilities of management and trade unions, and employment policies.

The original Code of Practice was intended to give industrial tribunals a yardstick by which to assess the merits of disputes; in the event it would appear that the tribunals have been pretty slavish in their devotion to the minutiae of the Code. All the more reason then that it be brought up to date and conceivably be made a bit more realistic.

CHANGES IN THE SOCIAL SECURITY AND SUPPLEMENTARY BENEFITS ACT

A source of much political and industrial controversy over the years has been the rule, dating from the original inception of a limited coverage of unemployment benefit in 1911, that:

> 'A workman who loses employment by reason of a stoppage of work which was due to a trade dispute at the factory, workshop or other premises at which he was employed shall be disqualified for receiving unemployment benefit so long as the strike or lockout continues.'

In consequence of this provision, process workers thrown out of work because of a strike of, say, maintenance men in the factory have had no entitlement to unemployment benefit.

In the Unemployment Insurance (No. 2) Act of 1924 there was a change in these provisions which was intended to be helpful to those who – like the process men cited above – found themselves disadvantaged by it. This Act provided that a man should not lose his right to unemployment benefit if he proved that:

- he was not *participating* in the trade dispute which caused the stoppage
- he was not *financing* the trade dispute which caused the stoppage
- he was not *directly interested* in the trade dispute which caused the stoppage
- he did not belong to a grade or class of workers, *some* members of whom were participating in the dispute (or financing it or directly interested in it).

This provision was narrowed in 1927 so that it applied only in respect of workers of the same grade or class employed at the premises where the stoppage occurred.

This situation was discussed extensively in the Donovan Report*

*See pages 246–56 of that Report.

as both the TUC and the CBI had views on changing the rules at that time.

The Royal Commission came to certain views on the matter some of which are relevant to the changes made in the Employment Protection Act. Thus the Commission wished to see the so-called 'grade or class' disqualification abolished and also the so-called 'financing' disqualification. The Employment Protection Act picks these views up and incorporates them into its provisions, adding also a clause which covers similar abolition in relation to the Supplementary Benefit Act.

We must now examine the net effect of these changes and try to estimate what effect they will have upon the pattern and conduct of trade disputes.

Before this change if, as Donovan noted, the process workers at a particular works went on strike on an issue which concerned them alone and one member out of a total of 100 maintenance workers struck in sympathy, the remaining 99 laid off (in consequence of the process workers' strike) would have lost their entitlement to unemployment benefit by occasion of application of the 'grade or class' rule. But otherwise they would have been entitled to benefit. With the abolition of these disqualifications, any employee who becomes unemployed by reason of a stoppage of work due to a strike at his place of employment will still have to prove that he was

- not participating in the dispute/and
- not directly interested in it

Further, the disqualification of 'financing' (e.g. belonging to the same union and contributing to a strike fund) is removed.

To sum up then, the Employment Protection Act, makes it easier for someone who is unemployed by reason of a strike or dispute in his factory to get unemployment benefit. It does this by removing the qualifications of 'grade or class' and 'financing' which have held sway since 1927. The Act leaves undisturbed the definition of 'place of employment' and the qualification of 'participating' and 'directly interested'. Clearly these changes will suit the unions; they mean that the chances of their having to pay lay-off pay to their members are reduced and that the possibility of pressure being exerted upon strikers by fellow workers is somewhat reduced.

In the related but separate matter of the change in the Supplementary Benefits Act, the effect of the changes i.e., the abolition of the grade or class and financing qualification, is that a man affected by a trade dispute will find it easier to claim supplementary benefit in lieu. Generally speaking, a man who is excluded from supplementary benefit for himself – but not for his wife and family – because he is affected by a trade dispute, will only be so excluded if he is *participating in* or *directly interested* in the dispute. If he is thrown out of work because of the strike, he can obtain supplementary benefit for himself as well as his family. His case, in fact, is parallel to the man who now qualifies for unemployment benefit.

These changes have long been sought by the TUC who would have liked to see entitlement to unemployment benefit and supplementary benefit in the circumstances of a trade dispute widened even further. The fact that the Act has extended those circumstances to some degree will have some effect, however marginal, upon the length, if not the incidence of strikes, and this effect will hardly be welcomed by employers, unless it helps to achieve the Act's stated aim of 'improving industrial relations'.

Chapter 7: Redundancy

Those sections of the Act which deal with obligations upon employers to consult with trade unions they recognise before making employees redundant, and to notify the Secretary of State in case of redundancy where more than ten people are to be made redundant over thirty days, have their origin in the EEC directive which deals with collective dismissal. The other countries in the EEC are, like ourselves, in the process of embodying the spirit and intention of the directive in their own legislation. It would be interesting to detail the differences which will thus arise between the legislation of one of the nine compared to another; we will however confine ourselves to describing our own.

Before going on to detail the provisions on handling redundancy it should be noted that the Employment Protection Act makes certain other changes in the law regarding redundancy and the operation of the Redundancy Fund.

First, the Act significantly amends the Redundancy Payments Act (1965) itself. Those changes and their effect upon employers' obligations are described, along with important changes concerning dismissal, in Chapter 2.

Secondly, the Act gives the Secretary of State the power to review each calendar year (or more often if he wishes) the limits placed upon calculation of redundancy pay by reason of the imposition of the maximum amount (presently £80) fixed for a week's pay. As we shall note in Chapter 9, the Secretary of State will be required, for this purpose, to put a draft order for approval to each House of Parliament. Further, if he decides not to vary this limit, he is required to say to each House why he has come to this conclusion. No doubt both Houses would be more than pleased to see inflation reduced to such a low level as to make such an annual review unnecessary.

Thirdly, the Act enables the Secretary of State to pay to employees certain debts which they are owed by an insolvent employer, and to do so out of the Redundancy Fund. This new drain upon the Redundancy Fund has been looked at in detail in Chapter 3.

However it should be noted here that such debts which include accrued holiday pay, any basic award of compensation for unfair dismissal and maternity pay, if paid out of the Redundancy Fund become, in effect, debts owed by the insolvent employer to the Secretary of State. If the Secretary of State succeeds in getting any of the money back he will pay it back into the Redundancy Fund. It is also of importance that such debts of the employee take priority over the claims of even debenture holders. In the event of the Secretary of State paying out from the Fund he in turn becomes a similarly high priority creditor. Nevertheless, the Government clearly expects a loss on the operation because the Act notes that these provisions are expected to add three million pounds a year to expenditure out of the Redundancy Fund.

CONSULTATION AND NOTIFICATION

This aspect of the Act involves changes of significance to employers as well as employees; as always in this Act a right for an employee is a new obligation and duty for his employer. Let us proceed however at this point to a major obligation which the Act places upon employers in connection with redundancy, namely the new procedures which an employer has to observe before he makes even one of his employees redundant, and which he has also to observe where he proposes to make what the EEC directive called 'mass redundancies'. As we shall see 'mass' in this connection means more than ten people over a period of not less than thirty days; evidently some masses are smaller than others.

Duty to consult with trade unions. First, then, the obligation to consult with trade unions over redundancy. The Act lays upon employers a general duty to consult representatives of an independent* trade union recognised by him which represents not only those employees proposed to be made redundant but also which represents employees of a similar description of those to be made redundant. This duty exists even if it is only one employee who is to be made redundant.

An example may be useful here. Suppose an employer wishes to make, say, six general labourers in a packing shed redundant. Let us

*'Independent trade union' has the meaning assigned by the Trade Union and Labour Relations Act of 1974, i.e. the union must be independent of an employer.

further suppose that all six are non-union members, but that the TGWU holds bargaining rights for general labourers in the packing shed. Despite the fact that only non-union men are involved, the TGWU has the right to be consulted about the timing, the method and the manner of the proposed redundancies. Only in a case where no independent trade union is recognised for that class of employee is the employer absolved from such consultation – he can hardly consult with that which he does not have (even though he may recognise independent trade unions for classes of employees not affected by the proposed redundancies). The intention of this part of the Employment Protection Act is to give trade unions notice of the intention of the employer to make certain employees redundant and thus not be in a position where the redundancies are sprung upon them. The employer has to give such notice 'at the earliest opportunity'; in the case of 'mass' redundancies (though the Act does not use that term). 'At the earliest opportunity' means the following:

– where the employer is proposing to dismiss as redundant ten or more employees in one establishment* within a period of thirty days or less, he must consult with the appropriate trade union *at least thirty days* before the first of these dismissals takes effect.

– where the employer is proposing to dismiss as redundant one hundred or more employees in one establishment within a period of ninety days or less, he must consult *at least ninety days* before the first dismissal takes effect.

In cases of redundancy – including those cases where only one employee is to be made redundant (as long as he belongs to a class of employee for which the employer recognises an independent trade union) – where such numbers are not involved the phrase 'at the earliest opportunity' will be decided, upon complaint from a trade union, by an industrial tribunal.

Exclusions from duty to consult. There will clearly be cases, such as sudden collapse of a market, insolvency of a major customer, major

*This term is not defined in the Act. Employers would be well advised to consider, where appropriate, whether or not 'establishment' here means the establishment affected by the redundancy situation.

technical mishap or even apprehension of trading when insolvent – when employers find themselves in circumstances which in the words of the Act 'render it not reasonably practicable' for the employer to comply with the obligation to consult trade unions. In these cases, it will be up to a tribunal, in the event of a complaint by a trade union of non-consultation, to decide whether or not the employer was justified (the onus of proof is upon him) in not consulting with the trade union because it was not reasonably practicable for him to do so. Even if he manages to do this, the employer must still prove that he took 'all such steps towards compliance' to discharge his duty to consult as were reasonably practicable. It should be noted by employers that if it *is possible* to comply with the obligation to consult, it will be very difficult to prove that it was not reasonably practicable to do so.

In any event, the employer will still have to reckon with industrial pressure to consult, or rather, pressure in the event of non-consultation. In militant situations where non-consultation might lead to a work-in or a sit-in, complying with the precise letter of the law here is probably the lesser of the worries faced by an employer.

In discharging his duty to consult, the Act lays it down that the employer shall *first* give certain information to trade union representatives who may be either lay representatives e.g., shop stewards and/or full-time trade union officials not employed by him. *Secondly*, once the trade union representatives have received (and presumably digested) this information, then the employer is required to consider any representations made by the trade union and reply to them. If the employer rejects any of these representations, then he is obliged to state his reasons for so doing.

So far so good, an employer might observe. He might very well then ask what are the penalties for not observing this duty to consult and as we shall see in the next section they are considerable. But first let us look at the information which he is required to disclose, remembering that it applies even in the case of only one or a few being made redundant, as well as the large figures quoted above.

(a) The employer is required to state *the reasons* why any employees have become redundant.

(b) He is required to list *the numbers and descriptions* of employees who he proposes to dismiss as redundant.

(c) He will be required to state also *the total number of employees of the same description* as those to be made redundant who are employed by him in the establishment affected by the proposed redundancies.*

(d) He will be required to give the *proposed method of selecting* the employees to be dismissed.

(e) He will be required to state the *proposed method of carrying out* the dismissals including the period over which the dismissals are to take effect.

Now clearly possession of this information will be of great advantage to trade unions and, it can be argued, to those whom they represent. Many employers will already have negotiated redundancy agreements which cover all or some of these requirements, and it is intended that all those who have not will be brought into line by these requirements. It is difficult to quarrel with such an intention; the aim of employers should be to arrive at timely and agreed procedures with trade unions to assure that as few employees as possible are declared redundant and to give every assistance to those made redundant to find other employment and so on. A difficulty at present, however, which the Act is silent about, is that the declaring of redundancies is increasingly resisted with an almost ideological fervour by unions and workers even to the extent of mounting sit-ins, plant occupations and the like. Many employers will no doubt feel that in such a climate the obligation to consult, which carries with it the obligation to give advance warning of redundancies, will have opposite effects to those intended by the Act. The provisions on consultation may in certain cases cause more trouble than heretofore. Many employers take the view that it is not because of ill-will that they delay the announcement of redundancies as long as possible; such a circumstance may imply that they have been unable to arrive at a redundancy agreement and that they know full well that all hell will break loose when the announcement is made. They fear not only short service workers bailing out rapidly to competitors, but also militant action which has as its end the prevention of redundancies – however necessary in the light of economic circumstances.

*Employers should note that the 'establishment affected by the proposed redundancies' is likely to mean also the unit for selection of redundant employees.

The Act, as we have said, does not concern itself with such matters. First the consultation, then the receiving and replying to representations and next, possibly, the penalties for not complying. And employers must note that the information being transmitted to trade unions is information on *proposed* redundancies.

Penalties for non-compliance. We have already glanced at the practical difficulties facing an employer when he decides that redundancies are necessary if he is to keep his business afloat and preserve the jobs of the remaining workers. At present a fine judgement has to be made, particularly so in the absence of a satisfactory redundancy agreement, on when and how to tell the unions of impending redundancies and it is not proposed here to add to the considerable volume of advice which presently exists on how to proceed in a redundancy situation. What we must now look at are the *legal* penalties for non-compliance.

A trade union may present a complaint to an industrial tribunal that an employer has dismissed an employee they represent (or one in a similar category to those they represent) as redundant, or proposes to do so, and they they have not, as the law requires, been consulted, or if consulted, not properly so.

In most circumstances, the employer will wish to defend himself against the charge. This he can do by trying to prove that he did indeed go through all the hoops prescribed by the Act – that he did consult at the earliest possible opportunity, that he did address himself to the appropriate trade union representative, that he did give the required information and so on. Alternatively he may try to prove that he could not go through all, or any, of the prescribed steps because it was not reasonably practicable for him to do so. The tribunal may, if it finds the trade union complaint well founded, make a declaration to that effect and additionally it may make what the Act refers to as a 'protective award'.

So, let us examine a case where a tribunal has found against an employer and has issued a declaration that he has not complied with this duty to consult. Let us further suppose that the tribunal decides to make a protective award. This award has the effect of requiring the employer to pay the employees made redundant (or proposed to be made redundant) a sum of money to compensate them in the light of the breach of duty by the employer. The details of the award may be summarised thus:

(a) The tribunal will normally* entertain a complaint of non-consultation only if it is presented to it before a proposed dismissal takes effect *or* before the end of the period of three months beginning with the date on which the dismissal takes effect.

(b) The period of the award shall start at the date on which the first of the dismissals takes effect *or* the date of the award; choice will be dictated by whichever of these dates is the earlier.

(c) The length of the period (which is of course crucial in determining the sum of money to be awarded) is up to the tribunal, which may bear in mind the degree of seriousness and the circumstances of the employer's default in not consulting.

(d) The protection period will not exceed:
   (i) any dismissal or dismissals (i.e., 1–9 people over whatever period) for redundancy: 28 days.
   (ii) '10 × 30 days' cases: 60 days.
   (iii) '100 × 90 days' cases: 90 days.

However, as we noted in (c) above the actual period specified in the protection award will be of such length as the tribunal shall determine to be 'just and equitable' in all the circumstances and having regard to the seriousness of non-compliance by the employer.

(e) Every employee affected by the protective award (which thus may include employees due to be made redundant as well as those actually made redundant) shall receive a week's pay for each week of the protected period. The employer may offset money paid under his contract of employment liabilities, i.e., money paid in lieu of notice or damages for breach of contract from the total sum assessed as the protective award.

(f) Further, if before redundancy has taken effect, the employer makes the affected employee an offer of his old job back or

*Which means there may be cases where the tribunal will decide that it was not reasonably practicable for the complaint to be presented within the time limit

a new job which is comparable as regards what is to be done for what terms (or alternatively an offer of other 'suitable'* employment) *and* the employee turns down an offer, then the employee forfeits his right to that bit of the protective award which overlaps the period when, had he accepted the offer, he would have been employed. It should also be noted that if the affected employee takes up a job offer and before or during the 'trial period' terminates that contract unreasonably, then again he is *not* entitled to that part of the protective award which overlaps the period when, if he had not terminated the contract, he would have been employed. (The trial is the period of four weeks beginning with the date on which the employee starts work under the contract as renewed, or the new contract, or such period as agreed between the employee and the employer.)

(g) Finally, the employee has recourse to a tribunal again if he wishes to complain that the employer has not paid the whole or part of the protective award to which he is entitled. The tribunal, if it finds the complaint well founded is obliged to issue an order against the employer.

The sequence of obligation to consult trade unions at the earliest opportunity, failure to do so resulting in possible complaint by the trade union to a tribunal, the issuing by the tribunal of a declaration that the employer has not consulted, the possible award of a 'protective award' and the complaint of non-payment of such an award by an employer to get another tribunal hearing, will strike employers as pretty onerous. It is. But that possible sequence of events will almost certainly oblige employers to plan their anticipated manpower requirements in a very careful way indeed. Some employers in the private sector will contend with justification that such a degree of precise anticipation of market requirements and consequent anticipation of manpower requirements is not possible. In these cases they will have to rely upon the tribunals' interpretation of the phrase 'reasonably practicable' in conjunction with a description of the circumstances. The burden upon small employers, the construction industry and certain distribution trades will be particularly great – to such a degree that the provisions may prove

*'Suitable' in context of Redundancy Payments Act.

to be unworkable. In this area, as in many others in the Act, we have the feeling that the drafters have the example of public employment very firmly in their sights. That area is one which is characterised by stability, by (alas) consistent growth, and even by detachment from the rude pressures of the market. The fact that the private sector has to be responsive to much sharper pressures has been largely ignored and employers in that sector will have to live with the results.

True, much of the potential damage can be obviated by careful forethought and sound administrative planning; it is to be hoped that in these difficult times private enterprise companies, large and small, will be able to afford the increased overheads that such planning requires. The checklist at the end of this chapter outlines steps in planning for the introduction of this new obligation and should be carefully studied.

Notifying the Secretary of State. We now turn to the final obligation placed upon employers by the section of the Act entitled 'Procedure for Handling Redundancies'. That is the duty of the employer to notify the Secretary of State of certain redundancies. The reasoning behind that duty is plain enough. The Department of Employment and particularly its associated agency, the Manpower Service Commission argue that advance warning of redundancies is required not only to plan the administrative action required for the immediate circumstances (already covered by the advance notice required by the Department of Employment by the Redundancy Payment Act) but also to enable the Government to take longer term action about training, industrial policy and so on.

The procedure is as follows:

(a) An employer proposing to dismiss as redundant:

(i) 10 or more employees at one establishment within a period of 30 days or less is obliged to notify the Secretary of State in writing of his proposal at least 60 days before the first of these dismissals takes effect.

(ii) 100 or more employees at one establishment within a period of 90 days or less, is obliged to notify the Secretary of State in writing of his proposal at least 90 days before the first of these dismissals takes effect.

(b) It will be appreciated from the foregoing section of this chapter that in these cases of 'mass' redundancy the employer is under a somewhat similar obligation to notify an independent trade union recognised by him for the class of employee affected within the same time limits. Thus there are dual obligations if the 'numbers' criteria above are achieved (there is only a duty to notify the trade union otherwise) to consult with a trade union or trade unions *and* to notify the Secretary of State. The Secretary of State will, in fact, require that a copy of the notice received by him will also be given by the employer to a relevant trade union(s).

(c) The notice to the Secretary of State will be in writing and will require the employer to state:

(i) the date on which consultation began with the recognised trade union (if applicable). This date theoretically could be after the date when the Secretary of State is notified, provided it is within the 60 or 90 day limit, but it will be appreciated that notice of the proposed redundancy has to be given simultaneously to a relevant union and to the Secretary of State. In most cases it would be advisable to plan for notification to the Secretary of State and the beginning of consultation with trade unions on the same date.

(ii) 'Such particulars as the Secretary of State may direct' will no doubt be outlined on a variety of the ubiquitous buff form. In any event, the Secretary of State has the right, once he has seen the data he originally requested to come back for more should he so direct.

(d) The employer may, in a manner analagous to defending his decision not to consult with trade unions, plead that it was not reasonably practicable for him to comply with this duty to notify the Secretary of State. In general terms, however, if an employer does not comply with this duty in the categories described above then the Secretary of State has one or other of two possible remedies. These are:

(i) the Secretary of State may take the employer before a magistrates' court where, if found guilty, he will be liable to a fine not exceeding £400. If the offence of non-notification has been committed with the consent or connivance of, or is attributable to any neglect on the part of a director, manager, company secretary or other officer of the company, then such a person or persons will also be liable to a fine as well as the company.

(ii) Alternatively the Secretary of State may decide, if the employer is entitled to a rebate on redundancy payments, to reduce that rebate by a proportion not exceeding one-tenth. If the Secretary of State chooses this remedy then the employer – rare circumstance – may appeal to an industrial tribunal to see if they agree with the Secretary of State's decision as to the reduction. If they do not the sum is adjusted accordingly. The time limit for such an appeal is three months.

So much for the procedures for notifying the Secretary of State. It should be noted that despite all the essentially delaying (and potentially costly) meaures forced upon the employer that his minute observance of them serves him for nothing in meeting his share of the cost of the redundancies finally achieved, to say nothing of the possibly greater tension generated by this public performance. Of course we may hope, in time, that these procedures will have the desired effect of enabling redundancies which may be in the national economic interest, to proceed smoothly with as little damage and disturbance to the company as well as providing practical help for those employees affected. We have a feeling that such a circumstance is liable to be arrived at somewhat earlier in the other eight Common Market countries affected by similar legislation before it happens here.

(A) REDUNDANCY PROCEDURES

Many companies have established procedures for handling redundancies and these may now need revision in the light of the new legal requirements described in this chapter. Where such procedures have been the subject of agreement with trade unions this is no less true, and companies will be well advised, having re-examined their agreements, to take the initiative in agreeing any necessary changes jointly with their unions – preferably before a redundancy situation occurs.

The following is intended to highlight in a general sense, the sort of issues often covered in redundancy procedures which will now need re-examination. It should not, however, be taken as a comprehensive list of every possible change to every type of agreement.

(a) Union(s) covered by agreement.

The new legal obligations to consult with trade unions, as mentioned earlier, apply to any independent trade union which the employer recognises as having bargaining rights for the employees who are to be made redundant or who are in a similar category as those to be made redundant. In establishments involving several recognised unions, therefore, it may be possible to reach agreement between all these unions on a common redundancy procedure since the new provisions will apply equally to all of them.

(b) Establishment covered by agreement.

Deciding and defining the establishment to be covered by the procedure is fundamental to the operation of the procedure in the new legal context. This is particularly so in a multi-plant or decentralised organisation. The ways in which the Act affects this issue have already been described but the following reminder may be useful:

- the *duration of statutory consultation* depends upon the total numbers of employees in the 'establishment' i.e., over 10 or over 100
- part of the information required to be given to appropriate unions is the total number of employees in the 'establishment' in question
- the appropriate union or unions must be informed of the criteria to be used for selecting redundant employees.

All of these factors are related. For example, a multi-plant employer may decide that one plant is the limit for redundancy and that the number of employees in that plant determines the appropriate length of statutory consultation. It would then follow, by deduction, that this plant is also the unit for selection of redundant employees. Conversely, if the 'establishment' means the entire organisation, then it would be argued that victims must be selected according to stated criteria from across the entire organisation. There would not seem to be any reason, however, why the employer should not use a wider area for purposes of redeployment than that used for selection.

It is therefore advisable for redundancy procedures to state specifically:

- the establishment within which a redundancy situation is contained, i.e., the unit for redundancy
- that the unit for selection is the same as above
- the establishment which may be used for purposes of redeployment which may be different from the above.

(c) Consultation.

Most redundancy procedure agreements provide for a specified period for consultation with trade union representatives. Statutory minimum periods, as we have seen, will now apply and employees will have to ensure that these periods at least meet these minimum periods. It should be noted, however, that these statutory periods apply *before the first dismissal takes effect*, whereas some agreements provide for consultation before redundancy notices *are issued.* In the latter case, employers may wish to ensure that the new periods are included or overlap the already agreed periods. In deciding when consultation should begin, the calculation should be marked backwards from the expiry of the shortest period of notice.

Lastly it should be pointed out that consultation with trade union representatives is a legal requirement even where less than ten employees are involved. Although no minimum period is specified in this case, it may be advisable to agree on the length of such a period and specify this in the procedure.

(d) Information.

The Act requires that certain specified information be given to the appropriate trade union representatives during the period of consultation. Such information includes details of the number of employees involved, the method of selecting employees and methods of carrying out dismissals. The type of information, and timing of disclosure should be stated in the agreement. The Act does not attempt to qualify either a method of selection for redundancy, or indeed any method of dismissal. It may be that ACAS will produce guidelines on these in one of its codes. If it does not, it may be wise to reach agreement with the trade unions on these points (where agreement does not already exist) and not wait for the uncompromising atmosphere of actual redundancy to resolve the matter.

(e) Criteria for selecting redundant employees.

For the reasons outlined above, employers will do well, where they have not already done so, to establish firm criteria for selecting redundant employees. This makes sense even where trade unions are not recognised, for interests of fairness and consistency; again also for the possible defending of claims of unfair selection for redundancy which are possible under the Trade Union and Labour Relations Act.

One of the commonest criterion specified in agreements is that of 'last in first out' but this has often proved undesirable in practice and does not always provide the employer with sufficient flexibility. Other possible criteria are:

voluntary redundancy (with employer reserving right to choose from volunteers)

part-timers

temporary staff

early retirement

those over retirement age.

Even when all the above criteria are used up they may still not make up the required number. The agreement may thus need to specify that the remainder will be selected with due regard to performance/efficiency, particular hardship, and length of service.

(B) CHECKLIST

1. Include detailed redundancy contingency plans in manpower planning.
2. Avoid redundancy situations by restriction/cessation of recruitment where long-term forecasting makes this possible.
3. Review existing redundancy procedures in light of changes considering:

   Unit of redundancy.

   Unit for redundancy selection.

   Redundancy selection criteria.

   Redeployment procedure.
4. Initiate re-negotiation of agreement with trade unions where appropriate.
5. Consider trying to establish common redundancy procedure by agreement where several unions are recognised in the same establishment.
6. Consider preparing managers' guide to handling redundancy situation including administrative drill, i.e., notification to Secretary of State (Department of Employment).
7. Ensure that possible trade union complaints of ineffective consultation to an industrial tribunal can be adequately defended by:

   Preparing and recording written notes of action taken, negotiations with trade unions, information disclosed, etc.
8. Where circumstances make the statutory consultation impossible, consult company's legal advisers before taking action.
9. Where this does not already exist, consider setting up a joint management/union team to handle certain aspects of redundancy, e.g., hardship cases, volunteers, redeployment and for general consultative purposes.

Chapter 8:

Statutory Wage Regulation

Although terms and conditions of employment in Britain are commonly settled by collective bargaining, statutory regulation has been applied to certain trades or industries where collective bargaining does not exist, or where it is not adequate for attaining a reasonable standard of remuneration for those working in the trades in question. Thus the state has set up Wages Councils to determine wages and conditions; in addition, through residual survivals of compulsory state arbitration like Section 8 of the Terms and Conditions of Employment Act 1959, the state in effect takes over when the trade unions through collective bargaining are unable to attain their ends; in this case the enforcing of conditions of agreements to which employers are not necessarily parties.

The philosophy of Section 8 of the 1959 Act and the Wages Council Act of the same year is much the same: if collective bargaining is unable to do its job properly then the state will, within modest limits, do it for it. Where the Acts differ is that Wages Councils, however permanent most of them might seem to those who are affected by their activities, are designed to be temporary substitutes for collective bargaining. When the latter shows evidence that it can do the job there is pressure for a Wages Council to disappear. There was no such air of impermanence over Section 8 activity. It was there to deal with employers who paid low wages even though they operated in industries already covered by collective bargaining. A Section 8 award was not a wage regulation order but an order that the employer comply with the collective agreement.

The Employment Protection Act does its bit, as we shall see, to widen this difference; the Act accelerates the possibility of collective bargaining taking over from Wages Council and makes the replacement provisions to Section 8 so attractive to trade unions as to make in certain circumstances, the possibility that they might prefer collective bargaining to the law even more remote.

The latter observations may be perceived by some to be somewhat exaggerated; in consequence we shall deal with the repeal of Section 8 of the 1959 Act first and see how the assertion can be substantiated.

COMPARABILITY CLAIMS AND THE LAW

The Employment Protection Act repeals section 8 of the Terms and Conditions of Employment Act 1959 and substitutes for it a potentially far-reaching set of provisions (Schedule 11). To understand the implications of this complex change it is necessary to see why it was that the 1959 Act got on to the Statute Book and how it has operated since. Section 8 of the 1959 Act was a relic of former arrangements for unilateral arbitration which were introduced, under the stress of war, in 1940. In that eventful year the Government approved the Conditions of Employment and National Arbitration Order and set up a National Arbitration Tribunal. This placed a prohibition on strikes and lock-outs and enabled compulsory arbitration to take their place. Arbitration could be invoked at the request of either the employers or the trade union. An award of the National Arbitration Tribunal became an implied term of the contracts of employment of the workers concerned.

In 1951 it was decided that since the prohibition of strikes and lock-outs could no longer be enforced (there had even in wartime been the spectacular case of defiance of this law in the affair of the Betteshanger Colliery) and in consequence the Industrial Disputes Order took the place of the 1940 order. Access to a re-named Industrial Disputes Tribunal was now only possible where no agreed arbitration arrangements existed or where existing voluntary arrangements had been used and failed. Breakaway unions and other unrecognised bodies were also prevented from making use of the arbitration machinery.

During the 50's a feeling developed among employers that the Industrial Disputes Tribunal procedure worked in a one-sided way and employers further contended that the availability of arbitration tended to undermine negotiations. As a result the tribunal was wound up in 1959 and all that was left of the former system was Section 8 of the 1959 Act. Under this section, representative trade unions (or employers' organisations) were able to get a judgement from the Industrial Court in cases where they were able to demonstrate that a particular employer was failing to observe terms and conditions of employment not less favourable than those established by agreement within an industry or section of industry. Employers' organisations had access to the procedure should they wish to stop rate cutting by other employers who would not, in practice, be members. This award, like the awards of the National Arbitration Tribunal and the Industrial Dispute Tribunal, became an implied term of the contract of employment of the workers concerned (and those subsequently employed). Awards could only be made if they referred to a relevant agreement. The phrase 'recognised terms or conditions' was used. It had to be established that (a) there was a relevant agreement or award; (b) that the parties to the relevant agreement (i.e., employers and trade unions) represented, respectively, a substantial proportion of the employers and the workers of the description to which the agreement related; and (c) that an employer was not observing those terms or conditions of employment mentioned in the agreement.

Mention in this brief historical survey, must also be made of the Fair Wages Resolutions which originated in 1891 with a resolution of the House of Commons; it was, in effect, the expression of the will of Parliament about the way the executive should safeguard the interests of workers employed by those to whom Government contracts were given. The latest expression of this will is the resolution of 1946. It states that a Government contractor is required to observe such terms and conditions as have been established for the trade or industry in the district by representative joint machinery of negotiation or by arbitration; or in the absence of such established standards, terms and conditions which are not less favourable than the *general level of wages, hours and conditions* observed by other employers where general circumstances in the trade or industry in which the contractor is engaged are similar. The italicised phrase is important in considering development in this field in the Employment Protection Act.

We now come to the changes brought in by the Employment Protection Act.

These may be summarised as follows:

(a) claims are to be reported to ACAS

(b) claims may not only be brought by unions (which must be 'independent') which represent a substantial proportion of the workers to which an agreement on 'recognised terms and conditions' (or minimum remuneration) applies but *also*

(c) by unions 'having members employed in the trade industry or section and in the district to which the claim relates' on the basis of the *'general level of terms and conditions'* observed by employers for comparable workers in the trade and workers where circumstances are 'broadly similar'.

(d) The Central Arbitration Committee will hear and determine the claim and will have regard to the whole of the terms and conditions observed by the employers. As before, these awards become part of the implied terms of the contract of employment of the workers who are the subject of the claim.

(e) If, in the course of determining a claim, it appears to the CAC that certain terms and conditions are applying specifically and exclusively to men or women then these shall *not* be taken into account in determining the general level of terms and conditions and that the CAC shall report its opinion to the Secretary of State (for possible action under the Equal Pay Act).

We now address ourselves to the perils of these procedures for employers. In general we can assume that few employers were likely to worry about a claim under Section 8 of the 1959 Act and there will be as few who will be worried about a claim by a union about the non-observance of recognised terms and conditions (particularly where there are minimum terms). All employers, however, have considerable cause to be alarmed by a claim from a union which need only have members in the trade and in the district about the non-observance of the general level of terms and conditions said to be enjoyed by comparable workers in the trade and in the district.

An example may help at this point. Suppose, say, the Association of Scientific, Technical and Managerial Staffs (ASTMS) has a few members who are supervisors in engineering factories in a particular district. Under the new provisions of the Act, that union (which has no national agreement on recognised terms and conditions at national level with the Engineering Employers' Federation (EEF) and thus would have had great difficulty in pursuing a claim under the old

Section 8 procedure), may now advance a claim based on the general level of terms and conditions for supervisors in those trades in the district. To establish a *general level* ACAS, say, would have to conduct a survey of all the supervisors in the engineering factories and arrive at an average. Claims would be considered against this average and if they proved to be below, an award would be made by the CAC. Such awards would, of course, change the average upwards and further consequential claims would be possible. The nightmare of a perpetual motion inflation machine becomes a possibility.

Furthermore the advantage to a union pursuing a claim under these provisions by the route of general conditions as opposed to recognised terms and conditions would be so great as to make some of them think that they would be much better off having no recognised terms at all – namely they would be better off having no national or district agreements. This outcome would be in line with the view of Donovan which stated that steps shall be taken 'to ensure that its (Section 8's) use does not hamper the development of company and plant bargaining'* but in a way which is so potentially inflationary as to be astounding.

Many employers are worried that unions in, for example, the engineering industry, may try to use these changed provisions to move up pay in small jobbing shops to the level of major car assembly plants. The National Union of Bank Employees (NUBE) has allegedly advertised its intention to use these procedures to get profit sharing schemes for its members from some clearing banks, because some others, where they have members, already have such schemes for NUBE members.

Employers will hardly be comforted by assurances from the Minister of State at the Department of Employment who stated in the House that this section of the Act was designed to clear up pockets of low pay. Only time (for this provision can hardly be introduced simultaneously with an incomes policy) – and the CAC – will tell if employers' fears are soundly based.

*Donovan Report, para. 275.

CHANGES IN WAGES COUNCILS' LEGISLATION

There were 46 Wages Councils in existence when the Employment Protection Act reached the Statute Book. They covered about 3½ million workers in the hotel and catering industry, retail distribution, road haulage, clothing manufacture, laundries, hairdressing and a number of minor industries. This large field is affected by the Wages Council Act of 1959 and that Act is amended extensively by the Employment Protection Act.

The Wages Council Act of 1959 consolidated the provisions of the Wages Council Acts of 1945 and 1948 and the Terms and Conditions of Employment Act 1959 so far as it related to Wages Councils. The Wages Council Act of 1959 was a consolidating act; it was designed, in combination with the Terms and Conditions Act, to bring into one system all the statutory wage regulating bodies outside agriculture. Writing about the 1959 Act in 1960, F. J. Bayliss noted:

> 'The problems which remain are not ones, which it is envisaged will need a legislative solution; future developments of the systems will come from the exercise of powers already provided.'

In 1968, however, the Donovan Commission stated:*

> 'Whatever may have been the case in the past, therefore, today many Wages Council are doing little to fulfill the aim of extending voluntary collective bargaining. In the main, areas of employment covered by Wages Council voluntary organisations remain too weak to support extensive machinery for voluntary negotiation and in some of the remaining Councils, the continued existence of statutory machinery seems to be a barrier to fully independent machinery.'

Donovan went on to make certain suggestions for Wages Councils, some of which coincide with changes brought about by the Employment Protection Act. For example, Donovan recommended that the Wages Councils Act 1959 should be amended to give the Secretary of State power to abolish any Wages Council, where he was satisfied,

*Donovan Report, para. 234.

on the application of a trade union which was representative of a substantial proportion of the workers concerned, that the wages and conditions of the work people would not be adversely affected if the Council in question was abolished.

The Employment Protection Act follows the recommendation to some extent by enabling the Secretary of State, by order, to change a Wages Council into a statutory joint industrial council; the abolition of a Wages Council can be accomplished as Donovan suggested, on application by a trade union nominated in relation to that Council, but the Secretary of State can also abolish it on application by an employers' association (or by joint application); the Secretary of State, can on his own initiative, also convert the Wages Council into a statutory joint industrial council, but in this case he must consult with the nominated* employers' associations and trade unions.

In future, Wages Councils will be able to make their own wage regulation orders instead of having to request the Secretary of State to do so. This should avoid the stalemate situation which often arose in the non-statutory wage freeze situation where the wages council recommended an increase but the Secretary of State refused to make the order. Also the jurisdiction of Wages Councils is extended to cover *all* terms and conditions of employment and not just wages and holiday pay as was the case under the 1959 Act.

Constraints laid upon the Secretary of State are that he is not empowered to 'convert', on a trade union or employer organisation application unless he has consulted every employers' association and trade union nominated in relation to the Council, and whether nominated or not, all trade unions and employers' organisations which represent a substantial proportion of workers and employers in the Wages Council's area. He is also required, before making an order, whether on application, or following consultations at his own initiative, to refer the question of whether he should do so to ACAS which has the duty of reporting on that question.

Conversion of a Wages Council to a statutory joint industrial council means that the independent members disappear. It should be noted that the powers of the Secretary of State as regards the nomination of the representative bodies which appoint the members of the Wages Council continue in relation to the two sides of the

*'Nominated' means persons who represent employers or workers in relation to whom the council operates; these persons are appointed by the Secretary of State.

new statutory joint industrial council. The joint industrial council consists only of employers' organisations and trade unions; whilst most existing joint industrial councils are based on the Ministry of Labour's 'Model Constitution and Functions'* they present considerable variety in structure, in their degreee of authority within the industry, and indeed in the nature and extent of their activities. Officers of the Department of Employment attend the meetings of a considerable number of joint industrial councils in the capacity of liaison officers. Most joint industrial councils negotiate wages and conditions but many also deal with matters such as unemployment, research and the collection of information on education, training, apprenticeship and so on.

Thus the decision to change a Wages Council to a statutory joint industrial council implies that the industry concerned has developed an organisation of employers and is sufficiently covered by trade union representation to allow collective bargaining (with certain safeguarding provisions at industry level) to be viable. The statutory joint industrial council need not confine itself to merely fixing minimum remuneration and holidays but can, if the bargainers so decide, run the whole gamut of matters nominally dealt with by collective bargaining at national level. The nanny figure of the Secretary of State is much less important (though it should be added that in practice the nanny-like powers of the Secretary of State in relation to Wages Councils have in the past been very sparingly exercised). The statutory joint industrial council will still have the responsibility of making orders which will regulate wages and conditions in the industry concerned, and there will be statutory enforcement in the same way as Wages Council orders.

In effect the Employment Protection Act allows for the possibility of setting up a statutory joint industrial council as a 'half way house' on the route to the ideal of collective bargaining in which statutory wage regulations plays no part. The Act also permits the Secretary of State to go one step further and abolish a statutory joint industrial council; this he can do by order on his own initiative (in which case he is obliged to consult the joint industrial council) or on application by the council itself. The Secretary of State would only so abolish a joint industrial council if he was of the view that 'adequate machinery' would be established and maintained for the

*See Appendix, page 209, *Industrial Relations* (Ministry of Labour), 1961 edition.

effective regulation of wages and other terms and conditions of the workers concerned. In this decision the Minister is obliged to take the opinion of ACAS, as in the decisions described previously, to change a Wages Council into a statutory joint industrial council. Once this final step is taken, however, the statutory powers of the joint industrial council cease, though its past orders remain in effect.

Donovan also recommended that the Wages Council Act be amended so as to enable the Secretary of State to exclude from the scope of any Wages Council any undertaking in which it was thought that there were satisfactory arrangements for voluntary collective bargaining for a substantial majority of its employees. The Act follows this recommendation, thus allowing the Secretary of State to make an order which excludes from a Wages Council any employers who are members of an employers' organisation which regulates remuneration or other terms and conditions by collective bargaining. This power means that where, for example a concern which is a member of an employers' organisation which collectively bargains on behalf of its members, the organisation may now also bargain for workers in that firm who were previously affected by minimum wage orders of a Wages Council. As Donovan observed:*

> 'There are numbers of large undertakings at present covered by Wages Councils, for example in distribution and catering, where it should be possible to develop satisfactory collective bargaining arrangements; and where this is done it is unnecessary and undesirable that they should be obliged also to observe statutory requirements in all their detail.'

Another Donovan recommendation in regard to Wages Councils was that Section 8 of the Terms and Conditions of Employment Act 1959 be amended so as to remove the prohibition of claims from workers in Wages Council industries. Donovan thought that wider effect would thus be given to voluntarily agreed standards and that the experience gained in the application of Section 8 procedure would be valuable when the time came to abolish statutory machinery and rely on voluntary means only. Now, as we have seen, Section 8 is itself repealed by the Employment Protection Act but is replaced by the potentially more extensive Schedule 11 of the Act. The Act thus follows Donovan when it states that claims under the new

*Para. 263.

procedure *may be* permitted in respect of workers covered by the Wages Council Act (and indeed by statutory joint industrial councils and by the Agricultural Wages Boards).

This section of the Act is however complicated by the fact that three types of comparison may be made on earnings in a Wages Council field of operation thus:

(a) 'recognised terms and conditions' (which may also mean minimum terms and conditions)

(b) 'general level of terms and conditions'

(c) 'the lowest current rate of remuneration'.

Comparisons with (a) and (b) are available in the way described in the previous parts of this chapter. Thus a claim may be reported to ACAS that an employer in a Wages Council area may be paying the statutory minimum prescribed (if he is not it is a matter for the Wages Inspectorate) but is not paying rates in line with those paid to workers in comparable employment in the trade or industry either generally *or in the district*. Alternatively a claim may be made to ACAS that the rates paid are not up to the general level of wages and conditions in the district. Finally, a claim may be made to ACAS that a worker who falls within the field of operation of a Wages Council (or statutory joint industrial council or Agricultural Wages Board) *where the union claiming is* a party to a collective agreement with an employers' organisation that covers a 'significant' number of establishments in the field of operation of the Wages Council. If we refer once again to Donovan (in para. 259) we note:

> 'There is, however, a good deal of evidence that statutory protection does not result in raising the pay of lower paid workers in relation to other workers.'

With the triple batteries described above there is good reason to think that this will no longer be the case. The combination of statutory remuneration and statutory comparisons should prove a potent, if expensive, weapon.

ACAS and statutory wages regulation. ACAS is given three specific duties in relation to Wages Councils and statutory joint industrial councils. It is required to perform those functions under the Wages Council Act 1959 which were formerly performed by a commission of enquiry, e.g., recommending the establishment of a Wages Council; it may suggest improvements in collective bargaining machinery for those workers who may be excluded from a Wages Council area; it can recommend the abolition (or variation) of a Wages Council, and issue reports on proposals to abolish statutory joint industrial councils. Secondly, disputes between the workers' and employers' sides of statutory joint industrial councils may be settled, upon request of the parties, by ACAS (using arbitration as a final step). And ACAS may also be asked by the Secretary of State to enquire into collectively bargained areas within the field of a Wages Council and decide whether or not to apply statutory minima to these areas, and enquire into the operation of the Wages Council Act 1959 and the provisions of that Act relating to Wages Councils and statutory joint industrial councils.

Information. The Secretary of State is empowered, in connection with the enforcement of an order specifying terms and conditions, to issue a notice to an employer covered by a Wages Council to furnish such information as may be specified in the notice to the order. In practice it is likely that such information will cover descriptions of employers, details of basic pay, premium and bonus earnings, holidays and other conditions. This power to obtain information is likely to be used to ensure that compliance with a wages order is widespread. Donovan complained (para. 257) that:

> 'There is, however, a regrettable lack of information about earnings of workers in Wages Council industries.'

The power of the Secretary of State to obtain information as described is clearly intended to make up for this deficiency.

Sanctions. The Employment Protection Act finally provides penalties for breaches of the Wages Council Act (and institutes penalties for disregard of the notice to supply information). Thus the maximum fine for failure to pay wages, or arrears, or to allow prescribed holidays is raised from £20 and £1 for each day on which the offence is continued after conviction, to *£100 and £5.* Again, payment of unlawful premiums, i.e., where an employer recovers a premium payment from an apprentice or a leaver to whom a wages order applies, was punishable by a fine of £20; it now becomes *£100.* Failing to produce documents or hindering officers, previously a maximum fine of £20, is now to be a *£100 maximum*. Producing false documents or producing false information moves from £20 to *£400.*

Evidently all of the changes described will have considerable effect upon employers in the various fields described. It is difficult to predict at what rate the changeover from Wages Council to statutory joint industrial councils to collective bargaining will proceed, however. Collective bargaining can only exist if employers are willing and able to set up organisations and procedures to deal with it; if they are neither willing nor able (say because of the paucity of union representation) it is difficult to see speedy widespread abolition of Wages Councils.

However, the other detailed changes will require study by the very many employers active within the field of Wages Councils. As is general in the Employment Protection Act a heavy responsibility is laid upon employers; we would hope for the good of many of the low paid workers employed in these areas that employers (and potential employers) are not over-discouraged by these additional responsibilities and penalties.

Part III:
The Institutions

Chapter 9:

Administrative and Legal Apparatus of the Act

In the previous chapters we have described the intentions and the likely effects of the Employment Protection Act; what effects they will have upon employees' rights and in consequence upon employers' obligations; what differences they will make to the process of collective bargaining, wage determination and disputes; and what employers will have to do to prepare themselves for the new situations arising from the new Act.

We now turn to a description of the functions and powers of the administration and legal apparatus which will be used in connection with the Employment Protection Act. Under the heading of 'administrative' we group the Secretary of State for Employment and the Department of Employment and the Advisory, Conciliation and Arbitration Service (including the important Central Arbitration Committee), and the Certification Officer. Under the heading 'legal' we shall describe the Industrial Tribunals, the Employment Appeal Tribunal, and, where necessary, the relevant divisions of the civil courts.

The function and operation of this apparatus is fundamental in achieving the purposes of the Employment Protection Act. Thus, we shall examine each of them, one by one. It is essential for an employer to understand in broad outline what each is supposed to do and how he may come into contact or collision with them.

THE SECRETARY OF STATE FOR EMPLOYMENT AND THE DEPARTMENT OF EMPLOYMENT

The old Ministry of Labour has had two further changes of name in recent years. First, to the Department of Employment and Productivity, and then to the Department of Employment. Under the present administration however, it has changed most radically, not this time in nomenclature but in organisational terms. The functions of employment exchanges and training have been hived off to the Manpower Services Commission, the Factory Inspectorate and relevant areas to the Health and Safety Commission and finally, by reasons of the passing of the Employment Protection Act, the functions of conciliation, arbitration and advice on industrial relations problems have been transferred to the Advisory, Conciliation and Arbitration Service run also by a tripartite council on much the same lines as Manpower Service and Health and Safety Commission. The rump of the Department is concerned with policy only and represents a rare example of a government department surrendering the control of its day-to-day operations to agencies made up of union and employer nominees and ministerial appointees. The policy rump of the Department is however vitally concerned with the exercise and operation of the powers of the Secretary of State in relation to the Employment Protection Act. These powers are:

Appointment of Certification Officer. After consultation with ACAS, the Secretary of State will appoint a Certification Officer whose principal duty will be to certify that a trade union applying to him is, or is not, an independent trade union.

Licensing of employment agencies. He will take over the licensing functions (outlined in the Employment Agencies Act 1973) from local authorities. He may thus refuse or revoke a licence; the agency affected may appeal, but only to the Secretary of State. In fact, given some of the provisions of the Act, concerning, e.g., temporary staff, employment agencies may become even more important to industry, in spite of their unpopularity with the trade union movement.

Jurisdiction of industrial tribunals. The Secretary of State may confer with the Lord Chancellor and the Secretary of State for Scotland to confer jurisdiction on industrial tribunals to deal with certain claims for breach of contract (*see Chapter 3*).

Duties and powers in relation to ACAS. He will appoint a full time Chairman to the ACAS Council, and will:

- consult with employers' organisations and organisations of workers he considers appropriate about the appointment of members of the Council;

- appoint up to three full time or part time Deputy Chairmen;
- approve of the appointment of the ACAS Secretary and of the members, and of the terms and conditions of service of officers (the latter in conjunction with the Minister for the Civil Service);
- receive in respect of each calendar year a report of the activities of the Advisory, Conciliation and Arbitration Service, the Central Arbitration Committee, and the Certification Officer; and receive copies of ACAS accounts.

Guaranteed work provisions. He can make an order excluding employees covered by a collective guaranteed work agreement from the operation of those sections of the Act giving individual employees a statutory right to a guaranteed week payment.

Phasing in the Act. The Secretary of State will determine when sections of the Act will come into operation (with the exception of those provisions relating to the setting up and functioning of the Employment Appeal Tribunal which will be the job of the Lord Chancellor). This means in effect that the Department of Employment, guided by political pressures and administration convenience will determine the bringing into law of the various sections. This is a most important power and employers will have to be vigilant about the appearance of these statutory instruments. In general terms, no rights and no obligations under any provision of the Employment Protection Act are operative until a Statutory Instrument has been laid. If we go on the precedent of the coming into force of the Industrial Relations Act, it is likely that the first process to be statutorily enacted will deal with setting up the various institutions described in this chapter. (*See Action Plan page 17*).

Wages Councils. He may make an order (subject to certain requirements in respect of relevant employers' organisations and trade unions and consultation with ACAS) to require a Wages Council to become a statutory joint industrial council (*see Chapter 8*).

He can by order abolish a statutory joint industrial council if he is of the view that adequate alternative machinery would then be established (in practice by collective bargaining) in its place. In this case, he may, but need not, be prompted by an application from the joint industrial council (JIC) itself, and he may, but need not, refer the question to ACAS.

By notice in writing, he can require an employer with the full co-operation of a Wages Council or a statutory joint industrial council, to furnish specified information and documents. Failure to comply would make the employer liable, upon summary conviction, to a fine not exceeding £100 (or £400 if making a false statement).

He can request ACAS to inquire into the development of agreements on wages and conditions within the field of a Wages Council or a statutory JIC, and to ascertain whether it is necessary for him to make an order on such terms and conditions under the Wages Council Act. He may also ask ACAS to enquire into the operation of the Wages Council Act; in either case he may ask ACAS to publish a report.

Redundancies. He will require an employer proposing to dismiss as redundant:

(a) 100 or more employees at one establishment within a period of 90 days or less or

(b) 10 or more employees at one establishment within a period of 30 days or less

to notify him in writing of his proposal. He will want at least 90 days' notice in the case of (a) and at least 60 days in the case of (b) – the notice relating to the period before the first of the dismissals takes effect. The Secretary of State will also be required to give a copy of the employer's notice to independent trade unions recognised in the establishment.

If an employer fails to give notice of proposed redundancies as described above, he can reduce the amount of rebate which the employer is entitled to under the Redundancy Payments Act by such proportions (not exceeding one tenth) as appears to him to be appropriate in the circumstances.

Alternatively, the Secretary of State may give his consent to proceed against an employer who fails to notify redundancies of the type described; an officer authorised by the Secretary of State may prosecute the offending employer before a magistrates' court; the employer will be liable upon summary conviction, to a fine not exceeding £400.

Insolvent employers. Upon application by an employee whose employer has become insolvent, the Secretary of State can pay the employee out of the Redundancy Fund. Similarly he may pay, on behalf of an employee or employees, occupational pension scheme payments not paid by an insolvent employer in respect of his employees; again, the Secretary of State will obtain the money from the Redundancy Fund. In each case he will take over the rights and remedies of the employee in respect of that debt (e.g. arrears of pay, holiday pay, or the basic award of compensation for unfair dismissal) and when and if he recovers that debt he reimburses the Redundancy Fund.

When he has paid contributions to an occupational pension scheme on behalf of an insolvent employer, the employer or any person having the custody or control of records (which may in practice be the officers of an insurance company) may be required to provide him with relevant documents. If such a request is not complied with, then the person responsible is liable, on summary conviction (i.e. in a magistrates' court) to a fine not exceeding £100, or if he makes a false statement to a fine not exceeding £400.

Financial limits. The Secretary of State is required each calendar year to review various financial limits upon payments by employers to employees. These include the limits of entitlement to guarantee payment (the Act now sets a maximum of £6 a day for not more than five days a quarter) the limits of entitlement to debts paid on behalf of an insolvent employer (presently £80 a week); the limits for calculation for the basic award for unfair dismissal (presently £80 a week); and the limits imposed by the Redundancy Payments Act as the amount of a week's pay.

In making such a review, the Secretary of State is required to take into account the general level of earnings prevailing in Great Britain at the time, the national economic situation as a whole, and such matters as he thinks relevant. If he varies any of these limits by reason of a review, he will propose a draft order for approval by each House of Parliament (and if he decides not so to vary any limit he is required to lay before each House of Parliament a report saying why not). He may also conduct reviews of these limits at periods of less than a calendar year – he is however, only required to do so once each calendar year.

The Secretary of State can, by order (and within specified limits) vary the redundancy numbers and timings specified above; vary the information required from the employer, and vary the notice period required. Such an order must be approved by a resolution of each House of Parliament. He may also vary or revoke the classes of employment (e.g. in unfair dismissal where the employer is the husband or wife of the employee, or where the employee ordinarily works outside Great Britain) excluded from various sections of the Act. Such orders must be approved by a resolution of each House of Parliament. He can also vary time limits attached to an employer's liability to pay employees who are suspended on certain medical grounds. Again, he may, by order, vary the requirements of the pay statement which every employee is now entitled to receive from his employer.

This list of the Secretary of State's duties and responsibilities in relation to the Employment Protection Act illustrates the key role which, in practice, the Department of Employment will play in operating the new law. Of great importance to the employer will be the power to vary (subject to approval by Parliament) the limits of entitlement to guarantee payment and the limits of calculation for the basic amount for unfair dismissal. We live in an inflationary age and there is explicit acceptance of that unpleasant fact in that the Secretary of State, while obliged to review the limits yearly, in the light of the general level of earnings prevailing in the country, may also conduct reviews which may change these limits at periods of less than a year.

Employers will be obliged, in making their estimates of cost in these matters, to anticipate that, given a continuation of inflation, these costs, like all others, will rise and they must plan accordingly. Even simple administrative arrangements will also be affected by the Secretary of State's power to vary the requirements of the pay statement given to employees.

Of even greater significance however, is his power of variation in the highly important matter of notification of redundancies. The full implications of this new requirement have been dealt with in Chapter 7, but suffice it to say here that the connection between the various time limits which may be varied by order, and trouble and expense for employers is very clear.

Employers in areas covered by Wages Councils will have to be alert also to the powers described in relation to these areas. The impatience of the trade unions with the low levels of pay prevailing in industries covered by Wages Councils will be strongly added to by the possibility of intervention by the Department of Employment and the likelihood is that there will be changes in these areas.

Finally, employers should note that in certain matters, notably the non-notification of redundancies, the Secretary of State (or his officers) has the power in the end to force him to a magistrates' court. Employers may very well feel, however, that such an indignity would be less unpleasant than the various other monetary penalties which may be levied upon him by industrial tribunals in the same event.

ADVISORY, CONCILIATION AND ARBITRATION SERVICE (ACAS)

In the same way as the Commission on Industrial Relations was given statutory recognition by the Industrial Relations Act after it was already operating, so the already established Advisory, Conciliation and Arbitration Service is given statutory recognition by the Employment Protection Act. The duties and responsibilities of ACAS will be described in this section, as will the other institutions, in terms of the Employment Protection Act. ACAS, however, is an institution of far reaching importance to the employer in general industrial relations terms. Thus, a description of the service and an outline of its significance for employers is given in Appendix 1, as well as the discussion of its powers throughout the book. In terms of the Act, ACAS is ubiquitous.

The Council of ACAS. The Act has a great deal to say (in Schedule 1, parts I and III) about the ACAS Council, but it is, as a whole, silent as to the relationship between the Council and ACAS. Thus we read, 'The Council shall determine its own procedure, including the quorum necessary for its meetings' and 'The functions of the Service . . . shall not be subject to direction of any kind from any Minister of the Crown as to the manner in which it is to exercise any of its functions . . .'

Inasmuch as the term 'Service' subsumes the Council of ACAS and its constituent services (Conciliation Service, Arbitration Service, manpower advisers, and so on) one is left in little doubt as to the overall responsibilities of ACAS. The Act says nothing, however, about the degree of control which the Council will exercise over the policy and operation of the constituent parts, which in practice will carry out the duties allotted to ACAS by the Act.

This point is by no means academic; employers will for example, be very interested when a draft of a Code on disclosure of information is completed, presumably by the staff of ACAS. It is a fair presumption that such a Code will have to be approved by ACAS Council. In the event of the three 'employer' members not approving the Code in whole or part, there could clearly be serious difficulties in the way of ACAS fulfilling at least one of its prescribed duties.

Duties. ACAS is given statutory being by the Act; it is referred to throughout the text of the Act as 'the Service'. It is given the general duty of promoting the improvement of industrial relations and of encouraging the extension of collective bargaining and the development and reform of collective bargaining machinery.

Appointments. ACAS will be controlled, as heretofore, by a Council, consisting of a full time Chairman, appointed as we have seen, by the Secretary of State, and nine other members. Of the nine, three have already been appointed by the Secretary of State in consultation with the CBI (though the Act refers only to the duty of the Secretary of State to appoint these three in consultation with 'such organisations representing employers as he considers appropriate') and three in consultation with the TUC ('such organisations of workers as he considers appropriate'). Such coyness as to organisations may be regarded as odd, in contrast to the Industrial Relations Act, which laid upon the Secretary of State a specific duty to consult the CBI and the TUC. While, conceivably, a splenetic Minister might decide the CBI was not an appropriate organisation to consult before appointing ACAS Council members, it is exceedingly unlikely he would choose, say, the Confederation of Employees' Organisations instead of the TUC.

The other three members may be appointed without consulting either. In addition, again as we have seen, the Secretary of State may (but need not) appoint up to three full time or part time Deputy Chairmen either from the numbers of the nine, or in addition to these members.

Conciliation. ACAS may, at the request of one or more parties to a trade dispute, offer the parties its assistance; this assistance will be by way of conciliation or what the Act describes as 'other means' which could mean informal advice by an outside person or body whose help or informal arbitration is obtained through ACAS. The former conciliation service of the Department of Employment will be used to execute this duty. This conciliation function of ACAS will continue in respect of unfair dismissals and will extend to the Sex Discrimination Act and other enactments.

Arbitration. ACAS may also at the request of one or more parties to a trade dispute, *and with the consent of all parties to that dispute*, refer the matter to arbitration *either* by one or more arbitrators appointed by ACAS (if more than one, ACAS will appoint one of them as Chairman) *or* by the Central Arbitration Committee. If, as is likely, 'consent of all parties' implies consent also to the form of arbitration, then consent of all parties, which of course includes the employer, will be required as to whether the arbitration mechanism chosen is a single arbitrator selected by ACAS, several chosen by ACAS, or recourse to the CAC. Again, by past practice, in the case of the first two, consent of all parties would be required for those individuals nominated as arbitrators. ACAS is not normally empowered to refer a dispute to arbitration until existing agreed procedures have been exhausted, unless all the parties agree to waive any stage of procedure to go to arbitration. Thus, if a trade union wished ACAS to set up an arbitration panel of three before an agreed procedure had been exhausted, ACAS could only do so if the employer agreed so to do.

Advice. ACAS is empowered either *on request* or *if it so desires*, to provide free of charge such advice as it thinks appropriate on *any* matter concerned with industrial relations or employment policies. The Act mentions the following as examples of those matters on which such advice might be proffered by ACAS:

- the organisation of workers or employers for the purpose of collective bargaining
- the recognition of trade unions by employers
- machinery for the negotiation of terms and conditions of employment and for joint consultation
- procedures for avoiding and settling disputes and workers' grievances
- facilities for officials of Trade Unions (including shop stewards)

- procedures relating to the dismissal of workers or other disciplinary matters
- manpower planning, labour turnover and absenteeism
- recruitment, promotion and vocational training of workers
- payment systems, including job evaluation and equal pay.

ACAS may also publish general, as opposed to specific, advice on any of these topics – or any other connected with industrial relations or employment policies.

Inquiry. In addition. ACAS may, if it thinks fit, inquire into *any* questions relating to industrial relations either generally or in relation to a particular industry or undertaking. Furthermore, ACAS may publish the results of such inquiries after taking into account the views of the parties appearing to ACAS to be concerned, and if it thinks such publication is desirable for the improvement of industrial relations.

Codes of Practice. The present Code of Good Industrial Relations Practice will stand, and as before, will not, in itself, have legal force. But again, as before, the Code will be admissable in evidence in any proceedings before an industrial tribunal or the CAC and if relevant, will be taken into account by these bodies.

ACAS is given the power to issue such Codes of Practice as it thinks fit for the purpose of promoting the improvement of industrial relations. In precise terms, however, the Act enjoins ACAS to produce two new Codes of Practice:

- on the disclosure of information by employers to trade union representatives for the purpose of collective bargaining
- on the question of time off for trade union officials.

If any part of these new Codes (or others) affects the present Code of Practice, then superseded parts of the Code will cease to have effect when the new Code comes into force (which will be accomplished by order of the Secretary of State, and subject to positive affirmation by both Houses of Parliament).

Recognition. As we saw in Chapter 4 one or more independent trade unions, but not an employer, may refer a recognition issue to ACAS which will then have a duty to try to conciliate on the dispute. If this is not possible then the issue may be referred to the CAC to be dealt with in a way described under the functions of CAC later in this chapter (and in chapter 4 itself).

Disclosure of information. While the main agency in cases of complaints by unions that an employer has not disclosed information needed for collective bargaining purposes is the CAC, ACAS may be given the task by the CAC of trying to settle the matter by conciliation. If it cannot, the CAC will then hear and determine the complaint.

Central Arbitration Committee. This important part of the 'Service' (ACAS) will consist of a chairman appointed by the Secretary of State after consultation with the Council of ACAS. The other members of the committee will be nominated by ACAS and appointed by the Secretary of State; they will include employers and trade unionists. The job of the CAC is a key one. It may be summarised as follows:

(a) *Arbitration in trade disputes.* Use of the CAC is one choice out of three (the others are an ACAS-nominated arbitrator, or a three man panel with Chairman appointed by ACAS) available to parties wishing to use arbitration as a means of settling disputes (*see Chapter 6*).

(b) *Disputes under Fair Wages Resolution.* The CAC takes over from the Industrial Arbitration Board in overseeing this measure, which is designed to ensure that employers engaged on Government contracts shall pay wages and observe other terms of employment not less favourable than those established generally by negotiation or by standard practice (*see Chapter 8*).

(c) *Disputes under the 'Extension of terms and conditions' provisions.* This extension is discussed fully in Chapter 8; CAC will hear and determine claims brought to the attention of ACAS by either unions or employers' organisations.

(d) *Complaints of failure to disclose information to an independent trade union.*

(i) CAC, if it finds such a complaint wholly or partly well founded will issue a declaration giving the date by which the employer must disclose the information specified in the declaration

(ii) CAC, on receipt of a further complaint that the employer has not disclosed the information he was directed to do by the time stated will issue a further declaration

(iii) CAC may also receive at the same time as the further claim detailed above (or after) a claim that the workers affected should have their contracts improved by the amount specified in the original claim. CAC is empowered to award the whole or part of the claim for improved terms or conditions or both. That award will be incorporated in the employees' contract of employment (*see Chapter 5*).

Complaints of employer's refusal to recognise a union

In the event of ACAS recommending the recognition of a union and, subject to the time limits described in Chapter 4, the employer refuses to comply, CAC may, in a fashion analogous to (d) arbitrate unilaterally on the terms and conditions specified in the union's claims.

In general terms the CAC, which clearly will be a most powerful institution, will take over the functions primarily exercised by the Industrial Arbitration Board which, following the passage of the Trade Union and Labour Relations Act 1974 itself took over some residual functions of the abolished NIRC.

CERTIFICATION OFFICER

The Registrar of Trade Unions and Employers' Associations was set up by the Industrial Relations Act 1971 and abolished by the Trade Union and Labour Relations Act 1974. Registration under the old Act involved a union accepting certain controlling standards in relation to its internal affairs in exchange for privileges and immunities which were not open to non-registered unions. In the event the TUC decided to make registration – or, more accurately, non-registration – the key area in which to demonstrate its resistance to the Industrial Relations Act.

In contrast the Employment Protection Act in setting up the post of a Certification Officer merely seeks to get done the jobs formerly performed by the Chief Registrar of Friendly Societies. It is an agency to certify that trade unions are truly independent. Under the 1974 Act the Registrar established a list of trade unions based on the following criteria:

- registered as a trade union prior to 30th September 1971
- immediately before the passing of the 1974 Act was affiliated to the TUC.

Unions on this list may apply to the Certification Officer for a certificate to show they are 'independent'. In order to be certified as an independent trade union, a union must satisfy the test laid down in the Trade Union and Labour Relations Act 1974, which provides that the union is independent, i.e. it is not under the domination or control of an employer or employers' association or is not liable (e.g. by the provision of financial support) to fall under such control. Independence is, of course, a key attribute in regard to many of the provisions of the Employment Protection Act. Thus only an *independent* union may apply to ACAS in a recognition dispute, or complain to the CAC that an employer has failed to disclose information needed for collective bargaining purposes. Any trade union aggrieved by the refusal of the Certification Officer to issue a certificate, or withdraw existing certification, may appeal to the Employment Appeal Tribunal. Such an appeal may, unlike the majority of appeals to the Employment Appeal Tribunal, be based on questions of fact *or* law.

INDUSTRIAL TRIBUNALS

These bodies are familiar enough already to employers; they adjudicate upon the contracts of employment, the redundancy payments, unfair dismissals and trade union membership. The Employment Protection Act adds th• following additional jurisdictions:

(i) Equal pay
(ii) Sex discrimination
(iii) Sick pay during suspension on medical grounds (*see Chapter 1*)
(iv) Maternity pay and return to work after pregnancy (*see Chapter 1*)
(v) Trade union membership and activity rights (*see Chapter 3*)
(vi) Time off for union and public duties (*see Chapter 1*)
(vii) Time off to look for new work (*see Chapter 1*)
(viii) Written reasons for dismissal (*see Chapter 2*)
(ix) Payment of wages and pensions contribution on insolvency (*see Chapter 3*)
(x) Itemised pay statements (*see Chapter 3*)
(xi) Protective awards – redundancy (*see Chapter 7*)
(xii) Breach of Contract of Employment (if ordered by Secretary of State – *see Chapter 3*).

EMPLOYMENT APPEAL TRIBUNAL

This tribunal will be set up not, as in the case of the other new institutions (i.e. ACAS, CAC, Certification Officer) by order of the Secretary of State for Employment but by order of the Lord Chancellor. This results from the composition of the members of the Employment Appeal Tribunal who will consist of judges of the High Court and the Court of Appeal nominated by the Lord Chancellor, one judge from the Court of Sessions nominated by the Lord President and other lay members nominated jointly by the Lord Chancellor and the Secretary of State.

EAT, unlike the National Industrial Relations Court established by the 1971 Industrial Relations Act, is not a court of first jurisdiction. It will hear appeals as points of law from industrial tribunals on the following matters:

(i) Redundancy Payments Act 1965
(ii) Equal Pay Act 1970
(iii) Contract of Employment Act 1972
(iv) Trade Union and Labour Relations Act 1974
(v) Sex Discrimination Act 1975
(vi) Employment Protection Act 1975

EAT will also deal with appeals both on law and on facts for decisions of the Certification Officer. It is interesting to note the presence of lay members. Since the majority of the appeals will only be possible on points of law it might seem odd, but it has been justified by the experience gained in the NIRC with lay members. In that now abolished court, lay members were useful in sorting out points of fact 'smuggled in' under guise of points of law and no doubt they will be equally useful and vigilant on EAT.

Appendix 1:

Advisory, Conciliation and Arbitration Service*

The Service is run by a Council consisting of a chairman and nine members – three nominated by the CBI (Confederation of British Industry), three by the TUC (Trades Union Congress), and three independent members. It is staffed by people with special experience of industrial relations.

The main functions of ACAS are to provide:

- Advice on industrial relations and the development of modern personnel practices
- Conciliation in industrial disputes
- Conciliation in certain disputes between individual employees and their employers
- Arbitration and mediation services
- Help in improving collective bargaining.

These services are closely related and in practice often interlink.

CONCILIATION

Where employers and trade unions cannot reach agreement by the normal process of collective bargaining, conciliation by an independent third party can make an important contribution to the settlement of disputes. ACAS may provide this conciliation if requested by the parties to a dispute. Or ACAS may offer conciliation (or give informal advice) where it considers this likely to be helpful. Trade unions should contact ACAS through their full-time officials.

Agreed procedures. ACAS takes account of any agreed procedure for dealing with disputes. Before responding to requests for conciliation, or in offering assistance to the parties, ACAS ascertains whether

*Reproduced by kind permission of the Advisory, Conciliation and Arbitration Service.

there is any agreed procedure and whether it has been fully used. It does not normally become involved where steps to resolve a dispute under an agreed procedure are still available to the parties. Where there is no agreed procedure ACAS expects the parties to have made an attempt to reach a settlement before seeking its help.

MEDIATION

Sometimes the parties in dispute may ask for the help of an independent third party to make recommendations for a settlement. Such help can best be described as mediation rather than conciliation. Subject to the same considerations concerning procedure agreements as apply to conciliation, ACAS has discretion to appoint a mediator from outside the Service at the joint request of the parties in dispute.

CONCILIATION IN INDIVIDUAL CASES

ACAS also conciliates in cases of alleged infringement of individual rights under industrial relations legislation, for example, the Contracts of Employment Act 1972, the Trade Union and Labour Relations Act 1974 and the Equal Pay Act 1970 which came into force on December 29 1975.

ACAS carries particular responsibility for dealing with complaints of alleged unfair dismissal. Specially trained conciliation officers are available at all ACAS offices to give advice and to try to settle such cases without the need for determination by a tribunal.

ARBITRATION

ACAS can make arrangements for arbitration in an industrial dispute if both parties want it. Arbitration of this kind is therefore voluntary; ACAS has no power to compel an unwilling party to go to arbitration.

As with conciliation, ACAS will normally take into account the existence of any agreed procedure for dealing with disputes when considering requests for arbitration. Arbitration awards are not legally binding, but since they arise from the parties' joint desire to secure a settlement such awards are normally implemented.

Before seeking arbitration efforts should be made to try to settle the dispute by conciliation.

Arbitration arranged by ACAS can take one of two forms: by a single arbitrator, or by a board of arbitration.

Single arbitration is undertaken by an independent person appointed to act alone without the assistance of side members. ACAS maintains a central list of experienced people from which such appointments are made. Hearings before a single arbitrator take place at a venue mutually convenient to the parties.

An ad hoc board of arbitration is temporary in character and is specially appointed to deal with a particular dispute. Under the chairmanship of an independent member appointed by ACAS it consists also of side members nominated by, or on behalf of, both parties – one or more representing the employer concerned and one or more representing the employees concerned. The parties are asked to agree that, where the board of arbitrators is not unanimous on what the award should be, the chairman shall have full powers of an umpire. The secretariat of such boards is provided by ACAS.

Requests for arbitration should be made to the appropriate ACAS Manager or to ACAS head office. Subsequent procedure is quick and informal. The arbitrators' fees and expenses are paid by ACAS; the parties' costs are limited to those incurred in presenting their case.

ADVISORY WORK

The broad aims of ACAS advisory work are to promote sound industrial relations and effective manpower and personnel policies. ACAS staff in each region are available to advise on the whole field of industrial relations. They diagnose any difficulties encountered and prescribe remedial action. Expert assistance is given on such matters as:

- Payment systems and conditions of employment
- Job evaluation
- Joint negotiating and consultative machinery
- Communications
- Grievance and disciplinary procedures
- Trade union recognition

- Absenteeism and labour turnover
- Recruitment, selection and training.

ACAS can provide assistance in several ways – by, for example:

- Short advisory visits to discuss specific problems.
- Surveys to diagnose the causes of more deep seated problems and recommend remedial action for discussion with management and with the trade union or employees' representatives concerned.
- A planned series of visits to assist the firm to adopt improved methods and practices. These may be initiated by management or recommended by ACAS and will involve discussions with management and usually trade union or employees' representatives.
- Helping to organise seminars and courses on industrial relations matters; and providing speakers.

IMPROVEMENT OF COLLECTIVE BARGAINING

ACAS has staff based at head office who carry out longer-term investigations to promote the improvement and extension of collective bargaining arrangements. These inquiries may involve wide-ranging studies of the way the whole industrial relations system is working in particular industries, companies or plants, or the inquiry might simply be concerned with a trade union recognition issue.

The aim is to suggest improvements by first establishing the facts, analysing them in detail and then recommending to the parties courses of action for dealing with the problems revealed. Discussions and consultations with the parties are maintained throughout the course of the inquiries so as to ensure that the situation is adequately interpreted and all views fully taken into account.

These inquiries generally require the agreement of the parties, though one party can initiate an investigation by an approach to ACAS. This type of inquiry is not generally appropriate for solving pay issues, though it might be used effectively to help in the reform of a payment system. An inquiry may be used in the development of a new procedure or used to establish the causes of conflict in an establishment. In the more complex industrial relations situations an inquiry can bring a detached and different perspective to the problems and therefore allow the parties to establish a new framework for their collective relationships.

A guiding principle in this type of work is that the most satisfactory progress is achieved by voluntary agreement. The ultimate value of any inquiry lies in persuading the parties to agree to the recommendations for future action.

ACAS ADDRESSES

HEAD OFFICE
Cleland House, Page Street, London SW1P 4ND
Tel: 01-222 4383

NORTHERN REGION
Wellbar House, Gallowgate,
Newcastle upon Tyne NE1 4TP 0632 27575

Cumbria	Tyne and Wear	Cleveland
Northumberland	Durham	

YORKSHIRE AND HUMBERSIDE REGION
City House, Leeds LS1 4JH 0532 38232

North Yorkshire	South Yorkshire	Humberside
West Yorkshire		

SOUTH EASTERN REGION
Hanway House, Red Lion Square
London WC1R 4NH 01-405 8454

Cambridgeshire	Bedfordshire	Kent
Norfolk	Hertfordshire	Hampshire (except Ringwood)
Suffolk	Essex	
Greater London	Isle of Ely	Isle of Wight
Oxfordshire	Berkshire	East Sussex
Buckinghamshire	Surrey	West Sussex

SOUTH WESTERN REGION
The Pithay, Bristol BS1 2NQ 0272 291071

Gloucestershire	Cornwall	Dorset
Avon	Devon	Ringwood
Wiltshire	Somerset	

MIDLANDS REGION
Fiveways House, Islington Row, Middleway,
Birmingham B15 1SG 021–643 8191

Derbyshire (except High Peak District)	Leicestershire	Herefordshire and Worcestershire
Nottinghamshire	Northamptonshire	West Midlands Metropolitan County
Lincolnshire	Shropshire	Warwickshire
	Staffordshire	

NORTH WESTERN REGION
Sunley Buildings, Piccadilly Plaza,
Manchester M60 7JS 061–832 9111

Lancashire	Greater Manchester	High Peak District of Derbyshire
Merseyside	Cheshire	

SCOTLAND
109 Waterloo Street, Glasgow G2 7BY 041–221 6852

WALES
2-4 Park Grove, Cardiff CF1 3QY 0222 45231

Appendix 2:
Industrial Tribunals*

Industrial tribunals
The tribunals are independent judicial bodies each with a legally qualified chairman appointed by the Lord Chancellor or in Scotland the Lord President. The two other members are selected from a list of persons appointed by the Secretary of State for Employment. The tribunals sit in different parts of the country as required.

IF YOU ARE AN APPLICANT

Making your application to the tribunal
1 Before you decide to make an application to a tribunal please read the leaflet and/or guide to the Act which gives you the right to apply. There is a full list of leaflets and guides for each Act in paragraph 31. It is particularly important for you to make sure that tribunals have the power to deal with the question you want to raise, and that you submit your application within the prescribed limits.

2 Your application should be in writing and give your own name and address, the name and address of the respondent(s) and the grounds on which you seek a tribunal decision. An application form designed to assist you can be obtained from any local employment office or unemployment benefit office. It should be sent to the Secretary of the Central Office of the Industrial Tribunals (England and Wales) if the matter to which your case relates arose in England or Wales. If the matter arose in Scotland you should send your application to the Secretary of the Central Office of the Industrial Tribunals (Scotland). The Addresses of these offices are given in paragraph 30.

3 A copy of your application will be sent to the respondent, that is the party whose action or decision you are disputing.* The respondent should, within 14 days of receiving the copy of your application, send in a written notice of appearance stating whether he intends to contest it and, if so, his grounds for contesting it. The Secretary of the Tribunals will send you a copy of this notice.

4 There are circumstances in which you may wish to withdraw your application before it reaches the tribunal hearing. The respondent, for example, may offer to settle your dispute with him without the need for a tribunal hearing, as described in paragraph 11, but if you decide to withdraw your application, you must put your request in writing to the Secretary of the Tribunals. Your application may then be formally dismissed. Without your written request for withdrawal the tribunal must hear the case in the normal way.
(Applicants should now turn to paragraph 11)

IF YOU ARE A RESPONDENT

Rights of respondents

5 If you are a respondent in a case, that is, if proceedings have been instituted against you for the determination of a question by an industrial tribunal, you have the right to contest the case before the tribunal.

Notice of the dispute

6 You will receive from the Secretary of the Tribunals a copy of the application in which the applicant's case is set out. You will also receive a notice of appearance on which you are asked to state whether you intend to resist the application and, if so, the grounds on which you propose to resist. If, in a redundancy case you are unable to pay the applicant because of financial difficulties you should include this fact in your statement of grounds on the notice of appearance.

Entering an appearance

7 To enter an appearance and to secure your right to take part in

*If you are an employer and appealing against the refusal of the Department of Employment to pay rebate from the Redundancy Fund the respondent will be the Secretary of State for Employment.

the proceedings, you should deliver the completed notice of appearance personally or by post to the Secretary of the Tribunals *at the address given in the notification to you* and within the required time limit of 14 days. The Secretary will then send a copy of your notice to the applicant.

8 If circumstances prevent you from sending the notice of appearance to the Secretary *within the 14 day time limit*, you should send it as soon as possible even though the time limit has expired, explaining the reasons for delay, as the tribunal has discretion to extend the time in which to enter an appearance.

9 Even if you decide not to resist the application you should complete the form accordingly and return it. This keeps open your right to appear at the hearing. If you do not enter a notice of appearance you will not be entitled to take part in tribunal proceedings, except to apply for an extension of time in which to enter an appearance. However the tribunal will be able to record a settlement reached voluntarily between yourself and the applicant. You may, however, be called as a witness by another person. You will in any case receive a copy of the notice giving the date and place of hearing and a copy of the tribunal's decision which may be given in your absence.

Settlement in Redundancy cases

10 In disputes about the entitlement to a redundancy payment, if, before the case reaches a tribunal, you wish to make a settlement with the applicant you should consult the Department of Employment before you reach any agreement with the applicant. Failure to do so may result in the refusal of rebate. This applies whether or not you have stated your intention to resist the application.

WHETHER YOU ARE AN APPLICANT OR A RESPONDENT

Conciliation

11 In cases which involve allegations of unfair dismissal the Secretary of the Tribunals will send copies of all statements to the conciliation officer of the Conciliation and Arbitration Service who will try to assist the parties to reach a settlement. Any information

given to the conciliation officer will not be admissible in evidence at a tribunal hearing without the consent of the party who gave the information to the conciliation officer.

Evidence

12 Where you are in dispute about matters of fact and there are other persons who have personal knowledge of these matters of fact, you should arrange for these parties to attend as witnesses to support your case. They may be required to give evidence on oath or affirmation. You should inform any witnesses you bring of the allowances which may be payable (see paragraphs 25 to 28).

13 In cases involving redundancy payment, unfair dismissal and written statements of the main terms of employment, it may assist the tribunal and shorten the proceedings if parties take with them any written statement that has been issued giving the particulars of the terms of employment. Where a tribunal upholds a complaint of unfair dismissal it must first consider the practicability of reinstatement or re-engagement. Both parties should therefore be prepared to give evidence on this; the tribunal may also need information on the pay, status and fringe benefits of the employee's last job in order to state the terms of a recommendation for reinstatement. It may also help in determining the amount of any redundancy payment due, or in assessing the appropriate level of compensation in a case of unfair dismissal where no recommendation is made for reinstatement or re-engagement, if any pay slips or other records of wages or salary, income tax forms, evidence of unemployment or other social security benefits, wages books and details of pension or superannuation schemes are brought to the hearing. Other evidence which may be of value to an industrial tribunal in assessing past and future loss following dismissal may include any anticipated pay increase in the old job, earnings from new employment (including earnings from self-employment), travelling expenses and removal expenses incurred in taking up a new job, loss of free accommodation and loss of a company car.

14 If you want any person to attend to give evidence as your witness you should arrange for him to come to the hearing. If he refuses, you can write to the tribunal asking for an order requiring him to attend. Similarly, if you want the other side to bring any

relevant documents to the hearing, you should ask them to do so. If they refuse you can apply to the tribunal for an order to require them to produce the documents.

15 You may also if necessary ask the tribunal to order that:
(i) the other party should furnish further particulars (which must be specified) about the case
(ii) documents in the possession of the other party should be produced for your inspection.

16 If the tribunal grants a witness order it is up to you to arrange to serve it on your witness and if necessary pay his fare which you can claim afterwards (see paragraph 26).

17 In addition to your observations on the application form or notice of appearance you may if you wish send written representations to the tribunal setting out the facts and arguments on which you intend to rely. These should be sent to the Secretary of the Tribunals so as to reach him at least seven days before the hearing. At the same time you should send a copy to the other party or parties.

Notice of hearing
18 You will be given at least 14 days' notice of the date, time and place of the hearing. In this connection it is very important to notify the Secretary of the Tribunals of any change of address, permanent or temporary, after submitting your application form or notice of appearance. If you are unable to attend you should inform the Secretary of the Tribunals *immediately*. If you have nominated a representative who agreed to act for you, the Notice will be sent only to him and he must ensure that you are notified of the time, date and place.

The hearing
19 The procedure is orderly but the atmosphere is informal and the tribunal assists a party during the presentation of his case. It is generally in your interests to attend the hearing since this will give you the opportunity to put your point of view in your own words, and to put the questions to the other party or parties or to their witnesses. You should not therefore, be put off from appearing merely because you are unaccustomed to presenting a case or for

fear of technicalities. In some cases a tribunal cannot reach a decision without the party's evidence given in person. Whilst every consideration is given to written representations, personal testimony is likely to carry greater weight.

20 At the hearing you can put your case in person or be represented by any person you wish to represent you for example, a representative of your trade union or employers' association. If you are not present or represented at the hearing, at the time and place notified, the tribunal may decide the case in your absence after considering any written representations which you may have submitted.

Private hearings

21 The hearing will normally be in public but you or any other party to the proceedings may apply to the tribunal for it to be held in private giving reasons. The tribunal is only likely to agree in exceptional circumstances.

Decision of the tribunal

22 The tribunal usually announces its decision and the reasons on which it is based at the close of the hearing. In every case the parties are sent a document setting out the decision and reasons. The payment of any monetary award ordered by the tribunal should be made direct to the party to whom it is awarded unless he is legally represented.

Appeals

23 You have the right to appeal against the tribunal's decision on a point of law to the High Court, or in Scotland the Court of Session. When you receive the copy of the decision you will also receive information about your right of appeal. If you decide to go ahead with an appeal you must give written notice to the High Court or, in Scotland, the Court of Session, within 42 days of the date on which the tribunal's decision is entered in the Register (as shown on the copy sent to you).

Review of a decision

24 The tribunal has the power, in limited circumstances, to review its decisions and to change or cancel them if it considers that they were wrong. If you think you have good grounds for asking for the

decision to be reviewed you may apply to the tribunal at the hearing immediately the chairman has given the decision or you may send to the Secretary of the Tribunals a written request, stating your reasons, within 14 days of the date on which the tribunal's decision is entered in the Register (as shown on the copy sent to you).

Payment of allowances

25 The following allowances may be paid to yourself, your witnesses and any person representing you (unless he is a full-time official of a workers' or employers' association, a barrister or a solicitor). Allowances to yourself and your witnesses may be refused if the tribunal decides that they were not reasonably justified.

Cost of travelling*

26 Cost of travelling by public transport (second class by railway) if the place of hearing is more than six miles from home or place of work within the United Kingdom or from the place of arrival in the United Kingdom. If a motor vehicle is used, the public transport rate of mileage allowance may be paid, plus ½p a mile for any passenger whose fare is admissible for payment. The rates are based on the cubic capacity of the motor vehicle and are as follows:

up to 150 cc	1.2p per mile
151 to 245 cc	1.8p per mile
246 to 500 cc	2.2p per mile
More than 500 cc	2.8p per mile

Subsistence allowances*

Period of absence from home	*Amount*
Less than 2½ hours	Nil
2½ hours but less than 5 hours	20p
5 hours but less than 10 hours	35p
10 hours or more	70p

Allowance additional to the above for an unavoidable night's absence £2.40, or, if the tribunal is held within 4 mile radius of Charing Cross, London, £2.70.

Loss of earnings*

28 Any earnings which are actually lost through absence from work

*The rates quoted were correct at the time of going to print but are subject to adjustment.

may be repaid up to £5.50 a day. Where the performance of remunerative work is merely advanced or deferred no compensation is payable. If you are self-employed and your attendance at a hearing has involved you in loss of earnings, you can receive an allowance equal to the amount you would have earned subject to a maximum of £5.50 a day. You should apply to the tribunal clerk, after the hearing, if you want to claim any of these allowances.

Award of costs or expenses

29 The tribunal will not order you to pay any part of the cost or expense of the hearing to the other party solely because you lose your case. But they have power to do so – and payment of any allowances may be refused – if they decide that you acted frivolously or vexatiously. They may also find you liable to make a payment to the other party if he suffers loss because a hearing has been postponed or adjourned at your request or because of some action or failure to act on your part.

General

30 Any communications about proceedings before an industrial tribunal, including notification of changes of address, should be made to the Secretary of the Central Office of the Industrial Tribunals at:

England & Wales	93 Ebury Bridge Road, London SW1W 8RE (Telephone 01–730 9161) or at the address subsequently notified to you.
Scotland	St. Andrew House, 141 West Nile Street, Glasgow G1 2RU. (Telephone 041–331 1601)

31 Matters which a tribunal may be asked to decide include those summarised in a note issued with the general application form (Form IT1 or IT1 (Scot)). This reads as follows:

"You can ask the tribunal to decide under the
Trade Union and Labour Relations Act 1974
on whether you were unfairly dismissed. You can find further

information about your rights in the booklet 'Unfair Dismissal' on whether you are entitled to belong to a trade union from which you have been excluded or expelled. See the 'Guide to the Trade Union and Labour Relations Act 1974'.

Redundancy Payments Act 1965
Questions about entitlement to, or amount of a redundancy payment or rebate. Further information is available in the booklet 'The Redundancy Payments Scheme'.

Contracts of Employment Act 1972
On the terms of your employment because you have not received a written statement of them or of any alteration to them, or because your written statement does not give sufficient information to meet the statutory requirements of the Act. For details see the booklet 'Guide to the Contracts of Employment Act 1972'."

32 The leaflets and guides referred to above deal with subjects which come within scope of the tribunals and are available free from any local employment office or unemployment benefit office. It is important for you to read carefully the leaflet or guide which deals with the type of case with which you are concerned. If you require full details of the procedure of the tribunal you should consult one of the following statutory instruments which you can buy from any branch of Her Majesty's Stationery Office or through any bookseller:

The Industrial Tribunals (Labour Relations) Regulations 1974
The Industrial Tribunals (Labour Relations) (Scotland) Regulations 1974

Appendix 3:

Extracts from 'The provision of information to employees'

FOREWORD BY THE PRESIDENT OF THE CBI – SIR RALPH BATEMAN

As you will know, one of the CBI's major objectives is to promote ways of enabling employers and employees to develop as effective a partnership as possible and of encouraging all those employed in the business to identify themselves more closely with its success.

For a number of years now the CBI has sought to develop policies in line with this objective and has taken a particular interest in the development of employee participation. Our initial conclusions were set out in the Companies Committee report *The Responsibilities of the British Public Company* in 1973. In the following year we published a more detailed discussion paper entitled *Employee Participation: CBI's Contribution to the Debate.*

Since then very many companies have been active in encouraging greater employee involvement in their affairs.

In this process our members have impressed upon us the need for detailed guidance on different aspects of this development. This need is made all the greater by current legislative proposals on the disclosure of information to employees.

It is for these reasons that the Council decided in May to issue practical advice for the consideration of members on the provision of information to employees. The Guidelines approved by Council are given in these pages. They will, we hope, be of real value to companies developing their policies in this field.

The Development of an Information Policy

WHAT KIND OF INFORMATION: GENERAL CONSIDERATIONS

5. Every company which has not already done so is recommended to take steps to develop a Company Information Policy dealing with the provision of information to employees. Because it is increasingly recognised that board level involvement is important for industrial relations, it is further recommended that the Information Policy be approved by the Board and continuously monitored at that level.

6. Many companies already have an Information Policy. There is, however, no one satisfactory model and whatever policy a company adopts must be developed in the light of its own situation and requirements, taking into account such facts as the size of the company, its structure and activities, characteristics of the work force, and existing negotiation and consultation arrangements.

7. In general terms, and subject to legal requirements, it must be the responsibility of senior management to decide what information should be made available to employees and how and when such information should be provided. It is recommended that in developing an Information Policy steps should be taken to discover what views employees at all levels, including middle management, hold on this matter since it will not always be possible for senior management to be fully aware of all needs. This may be done by the use of attitude surveys or questionnaires (in larger companies) or by formal and informal meetings with employees.

8. A piece-meal approach to the provision of information, or over-selectivity as to what is to be provided, may be self-defeating. Whilst it may be legitimate to draw distinctions between information which employees are required to have by law, or which they need in order to carry out their jobs efficiently, it would be a mistake to assume that employees are only interested in information of direct and personal relevance to their individual position and work in the company. Indeed, provision of information on this restricted basis would

probably not be in the company's own interest because it would not necessarily encourage employees to identify themselves with the company as a whole.

9. There are obviously important constraints on the provision of information. Management's freedom to be open is limited by the need to safeguard a company's competitive position, by the constraints of confidentiality and, in some cases, by sheer administrative difficulties.

10. The constraints which arise from competitive considerations are of absolutely vital importance. It is in the interests of everyone in a company that information should not be divulged if or at a time when to do so would have adverse consequences. Job security might well be jeopardised by inopportune or premature disclosure of information relating to mergers, takeovers or changes in location. Information about launching a new product or research and development should not be divulged if or when to do so would harm the company's prospects of sustaining or gaining a competitive edge.

11. Constraints of confidentiality may arise in a number of situations. Information should not be disclosed if it would be against national security; if it has been given in confidence to the employer; if it relates to an individual, unless he has consented to its disclosure; or if it relates to legal proceedings. Listed companies must also take Stock Exchange regulations into account.*

12. Administrative constraints may arise where the amount of work and expenditure involved in the compilation and assembly of information would be out of all reasonable proportion to the value such information would have.

13. There are powerful and legitimate constraints on the provision of information. The workforce and the public at large need to be made aware of them and to be sympathetic to them. They should not, however, be regarded as justifying a generally restrictive approach. Companies should bear in mind whether the consequences of failing to provide information might not be more damaging in industrial

*Stock Exchange regulations require that directors of listed companies should not divulge price-sensitive information in such a way as to place in a privileged position any person or class or category of persons outside the company and its advisers. Information should not be released in such a way that Stock Exchange transactions are entered into at prices which do not reflect the latest available information.

relations terms than the dangers of such information becoming common knowledge.

14. The latter consideration raises the question of whether, on some issues (for example, mergers, takeovers and closures) it might be wiser to disclose information to selected representatives of employees on a confidential basis than not to disclose it at all. Some bodies have taken the view that this should be possible providing that the representatives are prepared to agree to keep such information secret. However, a secrecy requirement could raise severe conflicts for the representatives involved and the choice of representatives to be informed could also be extremely difficult for the company. It is probably true that on this issue there can be no generally applicable guide. For this reason companies must decide such questions for themselves on the sole basis of their own circumstances and the nature of the matter involved.

WHAT KIND OF INFORMATION: A CHECKLIST

15. A Company Information Policy should contain a list of the types of information to be provided to employees. The following checklist is designed to assist companies in formulating this part of their policy.

16. Items marked with a black dot • are those whose disclosure to employees, to members of the company or to the public is already governed by law. The Appendix gives broad details, but since application varies as between different types of companies the legislation should itself be consulted.

17. Items marked with a star ★ are those where constraints of competitiveness or confidentiality may arise. See paragraphs 10 and 11.

Information about the Company as a whole

Organisation of the Company
Company activities •
Organisation of the company in relation to its subsidiaries or holding company (if any) •
Names of directors •
Information on shareholdings •

Unit structure
Management structure up to and including directors
Product range
General company objectives and policies

Finance
Turnover ●
Profits ●
Dividends ●
Losses ●
Liabilities ●
Total fixed assets ●
Details of directors' remuneration and emoluments ●
Details of chairman's remuneration and emoluments ●
Total wages and salary bill ●
Total labour costs as a percentage of operating costs ★
Administration costs as a percentage of total costs ★
Costs of materials ★

Competitive situation and productivity
Export performance ●
Details of main national and international competitors
Competitive possibilities
Orders, production and marketing situation
Production schedules ★
Work levels and standard performance levels ★
Labour costs per unit of output ★
Savings arising from increased productivity ★

Plans and prospects
Mergers and takeovers ★
Investment and expansion ★
Closures ★
Changes in location ★
Research and development ★
Product changes ★

Information relevant to employment*

Manpower
Average number employed ●
Numbers employed by sex, grade, occupation, departments, etc.
Labour turnover, redundancies and dismissals
Absenteeism
Redeployment, training and retraining
Days lost through disputes

Industrial Relations
Grievance procedures ●
Safety rules, policies and programmes ●
Health and welfare matters ●
Disciplinary and dismissal procedures
Trade Unions
Recruitment and promotion policies and methods
Induction programmes
Working rules
Social matters

Pay and conditions
Rates of pay ●
Notice periods ●
Hours of work ●
Holidays ●
Sick pay ●
Pension schemes ●
Results of negotiations affecting pay and conditions ●
Principles and structures of payment systems
Job evaluation systems
Holiday schedule arrangements
Profit sharing, stock option and saving schemes
Life assurance schemes
Average earnings

*Much of the information in this category has particular relevance to collective bargaining and it should be noted that under the Employment Protection Bill the Advisory, Conciliation and Arbitration Service would be required to formulate a Code of Practice on disclosure of information for this purpose.

Information relevant to the immediate work situation
Comparative departmental production figures
Comparative achievement of targets
Changes in the working environment
New appointments and changes in personnel
Any general changes affecting individuals in a working group.

18. The provision of financial information requires great care on the part of management. When such information is properly presented and explained, employees will gain a better understanding of their company's performance and prospects, of their own operating unit's situation and of how it and other units relate to the whole. However, in this area the dangers of misinterpretation are great. This is likely to be particularly true in relation to information about profits, firstly, because there may exist a 'political' credibility gap, and secondly, because a great deal of additional information may be required if a true and fair picture is to be given. The choice of definition of profit should take these factors into account. It is recommended that a definition should be adopted whereby net profit is always stated to indicate the final size of profit, but with surplus on trading used as the basis for the explanation of changes in operating results from period to period. This follows the presentation used in statutory accounts which commence with surplus on trading, and progress through profit before taxation to the final stage of net profit.

19. In general, care should be taken to ensure that the information employees receive is placed in a sufficiently broad context to make it understandable. It is of particular importance that employees should be able to understand their own company's activities in the overall context of the free enterprise system. This would involve, for example, an explanation of the role of profits and the way in which they are distributed, the importance of attracting investors and so on.

CONCLUSION

33. These Guidelines have been drawn up in the face of developing legal and social pressures on companies to provide information to employees and on the basis that it is in companies' own interests to respond positively if they are to secure from employees the co-

operation and commitment on which business success today depends.

34. In brief, it is recommended:

a. that in every company senior management, taking into account the views of employees at all levels, should formulate a Company Information Policy.

b. that this Policy should be approved and continuously monitored at the level of the Board.

c. that the objective of the Information Policy should be to provide employees with as much information as is relevant to their needs and wishes and which will assist them to identify with their company, paying due regard to constraints arising out of competitive requirements and confidentiality.

d. that each company should develop a communications system through which the Policy may be effectively implemented.

e. that each company should develop an appropriate training programme in order to assist the effective operation of the system.

35. It is hoped that these recommendations will be of real practical value to those companies now beginning to develop their policies in this area as well as to companies seeking to improve their existing arrangements.

Legislation Affecting the Provision of Information

COMPANIES ACTS

1. In addition to matters relating to the filing of accounts and the annual return, the present Companies Acts require directors in their annual report to 'deal with the state of the company's affairs and indicate the dividend they recommend and the amount they propose to carry to reserves' (Section 151(1) of the 1948 Act).

2. The **matters of a general nature** to be dealt with in the directors' report are set out in Section 16 of the 1967 Act and other matters are prescribed in Sections 17–20. The 1967 Act requirements (except for Section 17 – see item (i) below) apply to any report attached to the accounts of a company laid before it in general meeting in respect of

a financial year ending after the 27 January 1968. Broadly, the information required by Sections 16 to 20 is as follows:

a. the names of persons who have been directors during the financial year. S.16(1).

b. the principal activities of the company and its subsidiaries during the year and any significant changes therein. S.16(1).

c. particulars of any significant changes in the fixed assets of the company or any of its subsidiaries. S.16(1)(a).

d. details of shares and debentures issued during the year and the reasons for the issues. S.16(1)(b).

e. particulars of significant contracts with the company in which a director thereof has a direct or indirect interest. S.16(1)(c).

f. a statement of the arrangements to which the company is a party under which directors of the company are or were during the year enabled to acquire shares or debentures in the company or any other body corporate. S.16(1)(d).

g. information regarding directors' interests if existing at the end of the year, in shares or debentures of the company or its holding company's subsidiaries, etc. S.16(1)(e).

h. particulars of any other matters material for the appreciation of the state of the company's affairs. S.16(1)(f).

i. if the company has carried on business of two or more classes that differ substantially from each other, the turnover and the profit or loss attributable to each of those classes unless the company is neither a holding company nor a subsidiary company and its turnover does not exceed £250,000. S.17 and para 13A of the Schedule as amended by the Company (Accounts) Regulations 1971.

j. if the number of employees in the company exceeds 100, the average number employed during the year must be stated together with their aggregate remuneration. S.18.

k. particulars of contributions made by the company for political or charitable purposes (or both) if the total exceeds £50 in the year S.19.

l. if the company supplies goods, the value of goods exported during the year or a statement that no goods were exported unless:

(a) it is neither a holding company nor a subsidiary company and its turnover does not exceed £250,000 per year, or

(b) it is a subsidiary company but not a holding company and its turnover as shown in the accounts does not exceed £50,000 per year, or

(c) it is a holding company, whether or not it is also a subsidiary company, which presents consolidated accounts and its consolidated turnover as shown in the accounts does not exceed £50,000 per year. S.20 and para 13A of the Schedule as amended by the Companies (Accounts) Regulations 1971.

3. Every member of a company and any other person entitled to receive notice of a general meeting is entitled to a copy of the directors' report.

4. **Accounting information** must be sent to every member and debenture holder of the company prior to the general meeting whether or not they are entitled to attend the meeting. The following must be included:

a. a profit and loss account showing the profit or loss sustained during its latest trading year, together with a balance sheet showing the company's assets and liabilities on the date to which the above account is made up. S.148(1) and (2) of the 1948 Act.

b. every balance sheet must give a true and fair view of the state of the company's affairs at the end of its financial year and every profit and loss account must give a true and fair view of the company's profit and loss for that year. S.149(1) 1948.

c. the aggregate amount of directors' emoluments, the aggregate amount of directors' or past directors' pensions and the aggregate amount of any compensation to present and past directors in respect of loss of office during the period to which the accounts relate. Emoluments include the remuneration of all offices held by the directors under the company and its subsidiaries, such emoluments being shown separately from their remuneration as directors. Emoluments also include expense allowances, benefits in kind and contributions paid in respect of directors under pension schemes. S.196(1) 1948.

d. the emoluments (including expense allowances, benefits in kind and pension contributions) of the chairman of the board of directors. If any director(s) receives higher emoluments than the chairman, the emoluments of the highest one amongst them must be shown. S.6 1967 Act.

e. other directors' emoluments must be shown as an aggregate sum but the accounts must also show how many of them receive emoluments not exceeding £2,500 and how many receive emoluments within successive multiples of that figure. These particulars need not be given if the total amount of emoluments does not exceed £15,000 and the company is neither a holding company nor a subsidiary. S.6. 1967.

f. if directors have waived their rights to receive emoluments, the number of them who have done so must be stated, together with the total amount waived. S.7 1967 Act.

g. in the case of employees who received in excess of £10,000 there must be shown a scale dividing employees' emoluments into bands of £2,500, and the number of employees whose emoluments fell within each band of that scale. Employees who are directors of the company or who worked wholly or mainly outside the UK are exempt from these provisions. S.8 1967 Act.

h. the aggregate amount of loans made to officers of the company by the company itself or by its subsidiaries, or by other persons under guarantee or security given by the company and its subsidiaries during the period covered by the accounts and the global amount still outstanding of any such loans made before the period covered by the accounts. A loan need not be included if it was made by the company or its subsidiary in the ordinary course of its business, or if it was made to an employee of the company or its subsidiary and does not exceed £2,500. S.197(ii) 1948.

i. where at the end of its financial year a company has subsidiaries there must be shown in the accounts in respect of each subsidiary its name and place of incorporation and particulars of shareholdings in the subsidiary held by the company. S.3(i) 1967.

j. the names and places of incorporation of companies which are not subsidiaries whose shares it holds and particulars of those shares. S.4(i) 1967.

k. a subsidiary must include in its accounts the name and place of incorporation of the company's ultimate holding company.

5. Every company must also make up an **Annual Return** every calendar year and submit it to the Registrar at the Companies Register. The Return puts on public record in summary form facts of importance to anyone dealing with or concerned with the company. The following must be incuded:

a. situation of register of members and debenture holders.

b. particulars of company's share capital.

c. list of persons who, on the 14th day after the company's annual general meeting (the date to which the Return must be made up), are members of the company and of persons who have ceased to be members since the date of the last Return. The full list must be submitted every three years. In the interim only details of changes need be notified.

d. the company's debentures, mortgages and charges.

e. such particulars of directors and secretary as are required to be contained in the register of directors and secretaries maintained by each company at its registered office.

6. The following documents must be annexed to the annual return:

a. in the case of all limited companies and some unlimited companies, copies of the balance sheet and profit and loss acount, auditors' report on the accounts and the directors' report.

b. in the case of every private company certificates to the effect that (1) it has not issued any invitation to the public to subscribe for its shares or debentures (2) if it has more than 50 members, that the excess consists wholly of present or such past employees of the company as were in the company's employment when they became members of the company.

c. special disclosures for banking companies.

7. The documents to be lodged at the Registry have to be prepared in a statutory form.

CONTRACTS OF EMPLOYMENT ACT 1963

8. The Contracts of Employment Act 1963, as amended by the Industrial Relations Act 1971 and the Trade Union and Labour Relations Act 1974, requires broadly that except for excluded classes of employment every employee must be given a written statement, not later than 13 weeks after the commencement of his employment, setting out the main terms of his employment.

9. The written statement must give the names of the employer and employee and the date when the employment began. It must cover particulars of the terms of employment (at a date not more than a week before the statement is given to the employee) as follows:

a. the scale or rate of remuneration, or the method of calculating remuneration.

b. the intervals at which remuneration is paid.

c. any terms and conditions relating to hours of work, including any terms and conditions relating to normal working hours.

d. any terms and conditions relating to entitlement to holidays, including public holidays, and holiday pay sufficient to enable the employee's entitlement (including any entitlement to accrued holiday pay on the termination of employment) to be precisely calculated.

e. any terms and conditions relating to incapacity for work due to sickness or injury, including any provisions for sick pay.

f. any terms and conditions relating to pensions and pension schemes.

g. any terms and conditions relating to the length of notice the employee is obliged to give and entitled to receive. If the contract is for a fixed term, the date when the contract expires should be stated.

h. the person (specified either by name or by the postion he holds) to whom the employee can apply if he has a grievance about his employment and how the application should be made together with an explanation of the subsequent steps in any grievance procedure which may be available to the employee.

This list of particulars is taken from the Department of Employment Guide to the Act.

10. There is no prescribed form for giving this written information. The written statement itself need not set out all the required information. For all or any of the required particulars the written statement may refer the employee to a document or documents which he has reasonable opportunities to read or which are made reasonably accessible to him. If all or any of the terms of employment are set out in a collective agreement an employer may refer the employee to this. Other documents which may be used in the same way are works rules or handbooks, wages regulation orders, booklets about sick pay schemes or pension schemes and notices about such things as works holidays.

11. Written statements or reference documents must be kept up to date and changes in the terms must be notified.

HEALTH AND SAFETY AT WORK ETC ACT 1974

12. The Health and Safety at Work etc Act contains three Sections relating to the provision of information as follows:

a. under the general duty of every employer to ensure the health, safety and welfare at work of all his employees, the employer must provide such information, instruction, training and supervision as is necessary to ensure, so far as is reasonably practicable, the health or safety at work of his employees. S.2(2)(c). He must also provide copies of his safety and health policy with details of the organisation and arrangements for carrying out that policy. S.2(3).

b. Under the general duty of every employer to conduct his undertaking in such a way as to ensure that persons not in his employment who may be affected thereby are not thereby exposed to risks to their health or safety, it shall be the duty of every employer and every self-employed person in such cases as may be prescribed to give to persons not in their employment who may be affected by his activities prescribed information about such aspects of the way he conducts his undertaking as might affect their health or safety. S.3(3).

c. Section 16 of the Companies Act 1967 is to be amended to have included in matters to be disclosed in the directors' report information as may be prescribed about the arrangements in force in that year for securing the health, safety and welfare at work of employees of the company and its subsidiaries and for protecting other persons against risks to health or safety arising out of the activities at work of those employees.

d. duties are placed on Inspectors to disclose certain information to persons employed at premises to which the Act applies, to assist in keeping employees adequately informed about matters likely to affect their health, safety and welfare. This may include factual information obtained by the Inspector which relates to the premises or anything being done there and information with respect to any action which he has taken or proposes to take in, or in connection with, the premises. The Inspector is required to give the employer the same information as he gives to the employed persons.

12. Sub-paragraphs (b) and (c) above have not yet been implemented and await the issue of Regulations by the Secretary of State.

FINANCE ACT 1970: PENSION SCHEMES

13. In addition to the requirement under the Contracts of Employment Act that employees must be informed of any pension arrangements applicable to them, the Finance Act 1970 makes it a condition of approval of pension schemes for taxation purposes that every employee who is, or has a right to be, a member of the scheme has been given written particulars of all essential features of the scheme which concern him.

14. The revised Inland Revenue *Occupational Pension Schemes:*

Notes of Approval under the Finance Act 1970 as amended by the Finance Act 1971 details the particulars to be provided as follows:

a. main benefits, including those on normal retirement; on death whilst still in employment; on death after retirement; on withdrawal before retirement (including any optional methods of preserving benefits); on early retirement if provision is made for this; on any options at retirement, e.g. to surrender part of pension and take in lieu a dependant's pension or a lump sum.

b. basis of any contractual provisions for increasing pensions in payment or deferred benefits.

c. basis of contributions by members where the scheme is contributory, and administrative arrangements for collection. Any provisions governing the effect of temporary absence should be explained.

d. how the scheme as a whole is financed and the basis on which the employer contributes to it.

e. the legal constitution of the scheme including reference to the enactment under which it is or will be approved by the Inland Revenue.

f. particulars of the administrator, trustees and, where appropriate, any committee of management, including methods of appointment and replacement.

g. broad details of powers of amendment of benefits or rules and how changes can be effected.

15. It is worth noting that the Inland Revenue have based this list on a list contained in the booklet *Guidance on the Provision of Information to Members of Occupational Pension Schemes* published by CBI in July 1973.

Current legislative proposals

INDUSTRY BILL

16. Sections 20–24 of the Industry Bill contain provisions dealing with the disclosure of information by companies to a Minister and to representatives of relevant trade unions, i.e. independent trade unions recognised for bargaining purposes.

17. The disclosure requirements would apply to any person whom

the Minister designated as 'carrying on in the United Kingdom an undertaking engaged in manufacturing industry which makes a significant contribution to a sector of such industry important to the economy of the United Kingdom or to that of any substantial part of the United Kingdom or that a subsidiary of any person is carrying on such an undertaking.' S.20.

18. The minister would be empowered to require a designated person to furnish him with information as to any of the following:

a. the persons employed in the undertaking, or persons normally so employed.

b. the undertaking's capital expenditure.

c. fixed capital assets used in the undertaking.

d. any disposal or intended disposal of such assets.

e. any acquisition or intended acquisition of fixed capital assets for use in the undertaking.

f. the undertaking's output and productivity.

g. sales of the undertaking's products.

h. exports of those products.

i. sales of industrial property owned or used in connection with the undertaking, grants of rights in respect of such property and contracts for any such sales or grants. S.21.

19. The Minister's power to require information would include power to require an estimate as to 'the state of facts which will obtain at a future specified date or during a future specified period.' S.21(4).

20. The Minister would be required to order the information so furnished to be provided to a representative of each relevant union according to a list provided by the company, *unless:*

a. it would be undesirable in the national interest.

b. the person required to furnish the information could not do so without contravening a prohibition imposed by or under an enactment.

21. Section 23 of the Bill provides for the establishment of committees and an appeals procedure by which a designated person could apply for release from the duty to disclose. The Committee would direct release if it were satisfied that the information was given in confidence to the designee or if the disclosure of such information would be seriously prejudicial to the interests of the undertaking.

22. Information disclosed to the Minister under these provisions would not be disclosed without consent to any other person except to a government department, an appeals commmittee, the Man-

power Services Commission and its Agencies or in connection with criminal investigations and proceedings.

23. There are no confidentiality provisions relating to information supplied to a trade union.

THE EMPLOYMENT PROTECTION BILL

24. Section 17 of the Employment Protection Bill would impose on employers the duty to disclose to recognised trade unions, for the purpose of collective bargaining:

a. information without which the trade union representatives would be to a material extent impeded in carrying on collective bargaining, and

b. information which it would be good industrial relations practice to disclose. The ACAS would be required to issue a Code of Practice on disclosure which would be taken into account in deciding what information should be disclosed.

25. Employers would not be required to disclose information which would be against national security or was received in confidence or might be prejudicial to the employer's interests. Nor would the employer be required to compile or assemble information where the amount of work or expenditure involved would be out of reasonable proportion to the value of the information in collective bargaining. S.18.

26. In the event that an employer failed to disclose information, a trade union could complain to the Central Arbitration Committee. The complaint could be referred to conciliation but if this failed the CAC would decide on the complaint and, if it was well founded, make a declaration as to what information was relevant, the date on which the employer failed to disclose information and the period within which the employer ought to disclose the information. S.19.

27. A trade union could make a further complaint that the employer had continued to fail to disclose the information and could present a claim to the CAC in respect of terms and conditions of employment for the employees concerned. If the CAC decided the complaint was well founded it could make an award that the employer should observe either the terms and conditions of employment specified in the claim or other appropriate terms and conditions, which would become part of the contract of employment of the employees concerned. S.20.21.

28. Sections 89–96 of the Bill deal with the procedure for handling redundancies. If an employer proposed to make employees redundant and he recognised an independent trade union in respect of those employees he would be required to consult representatives of the trade union as soon as possible. For the purposes of this consultation the employer would be required to disclose to the union information as follows:

a. the reasons why any employees have become redundant.

b. the numbers and descriptions of employees whom it is proposed to dismiss as redundant.

c. the total number of employees of any such description employed by the employer at the establishment in question.

d. the proposed method of selecting the employees to be dismissed.

e. the proposed method of carrying out the dismissals, including the period over which the dismissals are to take effect.

29. Schedule 14 part II of the Bill would amend the Contracts of Employment Act by extending the written statement requirements to include details of any disciplinary rules and of a person to whom the employee may apply if he is dissatisfied with any disciplinary decision relating to him; and a specified date on which the employee's continuous period of employment began.

SOCIAL SECURITY PENSIONS BILL

30. Under provisions relating to the contracting out of pension schemes, Section 29 of the Bill provides for Regulations to be issued governing the procedure an employer must follow 'electing' to contract out. These would, *inter alia*, require an employer to give notice of his intentions in respect of any election:

a. to employees in any employment to which the scheme applies or to which it is proposed that it should apply.

b. to an independent trade union recognised for negotiating purposes in relation to those employees.

c. to trustees and managers of relevant schemes.

31. Section 29 also provides for Regulations to be made to require an employer, in connection with any such notice, to furnish such information as may be prescribed and to undertake such consultations as may be prescribed with any trade union as mentioned in (b) above.

Appendix 4:
Temporary Employment Subsidy

Hansard 7th August 1975

> **Mr. Jim Marshall** asked the Secretary of State for Employment how the temporary employment subsidy scheme is to be introduced before the Employment Protection Bill receives the Royal Assent.
>
> **Mr. Foot:** It is an accepted convention that once satisfactory provision is before the House in a Bill – or by way of a tabled amendment – the expenditure which it will authorise may be taken before enactment by virtue of the general powers in the Appropriations Act covering the relevant period. A supplementary estimate will be presented in due course. In the meantime recourse will be made to the Contingencies Fund.

The Employment Protection Act enables general provisions, of which the Temporary Employment Subsidy (TES) is an example, to be introduced. The *Hansard* quotation shows TES could come into operation before the Employment Protection Act was passed.

What is Temporary Employment Subsidy (TES)?

The TES scheme is one of several measures drawn up by the Government to asist in combating rising unemployment. It is intended to alleviate some of the effects of high unemployment in the worst-hit areas by providing for a short-term subsidy to be offered to companies in Assisted Areas (i.e., Development Areas, Special Development Areas and Intermediate Areas) who are prepared in consultation with the trade union(s) concerned to defer impending redundancies affecting 50 or more workers.

The scheme is entirely within the discretion of the Secretary of State for Employment.

Employers' participation is purely voluntary; they must judge, in the light of their individual circumstances, whether the scheme is likely to be beneficial.

What are the aims of the scheme?

It is hoped that, at the end of the period of payment of the subsidy the company's business would have recovered sufficiently to enable the company to retain the workers concerned in employment. In other cases, however, the period of subsidy might be used to facilitate the redeployment of the workers (e.g., through retraining) into other jobs, whether within the company or elsewhere.

In appropriate cases (e.g., where there is no prospect of continued employment for a particular redundant worker at the end of the period of subsidy), employers may be asked whether they would like to consider the possibility of excluding the worker from the redundancy group covered by the scheme, and instead offer him training under the Training Before Redundancy (TBR) scheme which is run by the Training Services Agency. Similarly, a worker who chooses not to have his notice of dismissal rescinded would be eligible for the TBR scheme.

Who is covered by the TES scheme?

All employment in the private sector of industry and commerce in Assisted Areas is covered by the scheme, and employees in all occupations may be included, provided that in the current period they work on average not less than 21 hours per week.

Absences, such as sick leave, holidays or absence without leave will be ignored.

TES is *not* applicable in Northern Ireland.

How long will the scheme last?

TES is a temporary measure lasting one year only (i.e. applications must be received within one year of the date of inception of the scheme).

What are the conditions for the receipt of TES?

1 That the redundancies occur in an establishment wholly within an Assisted Area. (A worker's place of residence is immaterial, but workers whose place of employment is outside the Assisted Area may not be included.)

2 That there is evidence of a decision in good faith to declare an impending redundancy affecting 50 or more workers, and that the Department of Employment is satisfied that the company genuinely intends to dismiss the workers as redundant if it does not receive the subsidy. An impending redundancy for TES purposes is a proposal to dismiss 50 or more employees at one establishment within a period of 30 days or less or 100 or more workers over a period of 90 days or less (the period over which dismissals are planned). Additionally, the first of the dismissals must be scheduled to take effect not more than 90 days from the date of application for TES.

3 That the company has begun consultations about the redundancies with the trade union(s) concerned, and that the application form is jointly signed by the company and the trade union representative(s).

4 That, in the opinion of the Secretary of State for Employment, the company is not insolvent or about to become insolvent. It is a specific condition of the scheme that payments will not be made or continued if the declarations of solvency are not signed when required, or the Department of Employment is otherwise not satisfied about the continued solvency of the recipient. *It must be understood that the scheme is intended to encourage the retention of workers in employment,* not *to enable companies to continue to trade when otherwise they would be unable to and that the question whether companies can properly continue trading after the subsidy is paid will be entirely a matter for the discretion of those companies.*

5 That there are reasonable prospects that, with the assistance of the subsidy, the workers concerned would be kept in employment for the period of the subsidy, and either retained in the firm or re-deployed more effectively thereafter.

6 That the provisions of the Government White Paper, *The Attack on Inflation* regarding the pay limit are adhered to. Subsidy will not be available to companies which exceed the pay limit.

How much is the subsidy?

The amount of subsidy payable in respect of each full-time worker whose redundancy is deferred is a flat rate of £10 per week, regardless of age or sex. Payment in respect of workers working, on average in the current period, 21 hours or more per week, but less than full-time, will be determined on a pro rata basis.

For how long will it be paid?

In the first instance, payments will be authorised for a period of three months, subject to affirmation at the end of each month by the employer that the general situation has not changed. Subsidy will be paid monthly in arrears, and the amounts payable will be based on the number of redundant jobs certified by the employer as continuing to be maintained.

At the end of three months the employer may re-apply for subsidy and, providing the qualifying conditions continue to be satisfied, payments calculated on the same basis as above may be authorised for a further three months, *giving an overall maximum of six months.*

In what circumstances can payments cease?

Payments may be terminated at the monthly stage, or otherwise, where, in the opinion of the Secretary of State for Employment, the conditions or purposes of the scheme are no longer satisfied.

The employer is free at any time to terminate his participation in the scheme, providing proper written notification is given to the Department of Employment.

Can subsequent application be made?

No increase is payable under the scheme when a further group of workers numbering less than 50 becomes liable to redundancy. An employer may however, make further application for assistance in respect of a subsequent redundancy affecting a different group of 50 or more workers.

How should the initial application be made?

Application form TES 1 may be obtained from the Department of Employment's Regional Offices or Unemployment Benefit Offices, or from any Employment Office of the Employment Service Agency.

The completed application form should be sent to the TES designated officer at the Regional Office which administers the area in which the redundancy occurs.

Regional Offices covering Assisted Areas

Northern Regional Office

93A Grey Street, Newcastle upon Tyne NE1 6HE

Newcastle 26011 (STD code 0632)

Yorkshire and Humberside Regional Office
City House, Leeds LS1 4JH
Leeds 38232 (STD code 0532)

South Western Regional Office
The Pithay, Bristol BS1 2NQ
Bristol 291071 (STD code 0272)

Midlands Regional Office
2 Duchess Place, Hagley Road,
Birmingham B16 8NS
021–455 7111

North Western Regional Office
Sunley Building, Piccadilly Plaza, Manchester M60 7JS
061–832 9111

Office for Scotland
Stuart House, Semple Street, Edinburgh EHS 8YX
031–229 2433

Office for Wales
Dominion House, Queen Street, Cardiff CF1 4NS
Cardiff 32961 (STD code 0222)

Regional Office in non-Assisted Areas

South East Regional Office
Hanway House, 27–31 Red Lion Square
London WC1R 4NH
01–405 8454

Index